Understanding Mathematics

Understanding Mathematics

Second Edition | KEITH GREGSON

The Johns Hopkins University Press | *Baltimore*

© 2007, 2010 Keith Gregson

First published by Nottingham University Press represented by Cathy Miller Foreign Rights Agency, London, England

North American Edition © 2010 The Johns Hopkins University Press
All rights reserved. Published 2010
Printed in the United States of America on acid-free paper
9 8 7 6 5 4 3 2 1

The Johns Hopkins University Press
2715 North Charles Street
Baltimore, Maryland 21218-4363
www.press.jhu.edu

ISBN 13: 978-0-8018-9701-6 (hc)
ISBN 10: 0-8018-9701-7 (hc)
ISBN 13: 978-0-8018-9702-3 (pbk)
ISBN 10: 0-8018-9702-5 (pbk)

Library of Congress Control Number: 2009943463

A catalog record for this book is available from the British Library.

Special discounts are available for bulk purchases of this book. For more information, please contact Special Sales at 410-516-6936 or specialsales@press.jhu.edu.

The Johns Hopkins University Press uses environmentally friendly book materials, including recycled text paper that is composed of at least 30 percent post-consumer waste, whenever possible. All of our book papers are acid-free, and our jackets and covers are printed on paper with recycled content.

To Eileen, Emma, and Michael

Never measure the height of a mountain until you have reached the top. Then you will see how low it was.
— From the diary of Dag Hammarskjold (1905–1961), Swedish diplomat and political economist

Contents

Preface

This book is written for those of you who are struggling with mathematics, either as first-year undergraduates taking math as a subsidiary to a science course, or students taking AP courses or working for S- and A-levels. It is not a comprehensive syllabus text—there are many such books and software courses.

Some students take to mathematics easily, but others (the majority) have difficulty. This difficulty is generally the result of having a hard time throughout earlier education. I know that many of you have been told repeatedly that mathematics is hard. Much of it is not—but mastering it does involve a different approach; memorizing a few formulae is not the answer. Success comes from reading, several times if necessary, together with thinking, until the mathematics is understood. Sometimes understanding doesn't come easily. You will need to be able to seek help: ask your fellow students, or tutors, or whomever. This takes effort—*from you*—but what you put in will pay great dividends, so don't give up without a fight! Attitude of mind is important; as succinctly put by Henry Ford, "Whether you believe you can do a thing or not, you are right."

The book is made up from notes that I have given to first-year undergraduate students in biosciences, though it could be equally useful to students in other areas of science and engineering. Its aim is to encourage the student to appreciate how mathematical tools are derived from a few simple assumptions and definitions, and thereby to discourage the examination habit of memorize, recall, and forget. The emphasis is to encourage students to think about a few topics in order to help them develop an understanding of the underlying principles of mathematics.

Read the book steadily, a little at a time, working through the proofs and examples until you are confident that you understand them. Try the exercises, but don't spend more than 20 minutes on any of them. If you can't solve them, try to identify the difficulty so that you can ask for help, and then ask!

Many mathematical formulae can be found from first principles without too much difficulty, and I make no apology for including some of the derivations here. I do not expect you to commit them to memory. They are included here to promote understanding and confidence, though you will be surprised how much you can remember once you have followed the logic. Difficult formulae are best looked up, but it is helpful to have sufficient understanding to know where to start looking. It is also important in this Internet age to know that what you are looking at has some authority.

I have deliberately avoided the rigorously inflexible approach which is essential to a pure mathematics readership. Rather, I have attempted to develop the spirit of the subject without examining all the details. As in life, it is often better not to question every assumption, lest doing so detract from the more important (and fun) moments. As new topics are introduced I have gradually reduced the verbal description, in line with the (I hope) growing ability of the reader to handle the notation and brevity.

The idea for this book came from some of my students. I hope that the result will both please them and be useful to others. I thank them for their belief.

I must also thank my wife for helping me find the time to write the book (by taking on all the other things that I should have been doing!), and my family simply for being themselves—I count myself lucky.

Last, but not least, I would like to express my gratitude to the staff at the Johns Hopkins and Nottingham University Presses, both for their help and for the manner in which it has been given.

Understanding Mathematics

Chapter 1

Fundamentals

1.1 Why Mathematics?

It is interesting to see the approach that newspapers, even the so-called informed and respectable press, have toward the current craze for Sudoku. Almost without exception they will tell you that Sudoku problems can be solved without recourse to mathematics. They qualify this statement with the further claim that the puzzles may be solved merely with reasoning and logic. They clearly do not understand what mathematics is about—so if you are having difficulty, you are not alone!

It is true that Sudoku problems are not numerical—you could just as easily substitute a,b,c, ...for 1,2,3, ...—but mathematics involves much more than arithmetic: logic and reasoning, for example.

In the case of Sudoku, newspapers are all perpetuating the myth that mathematics is hard and should be avoided at all costs. That some mathematics is difficult is undeniable, but much of it is not. So where does the myth come from?

Some of it arises because we (most of us) do not like to be fenced in by rules and regulations—and mathematics depends upon rules, though there are surprisingly few of them. You can't simply scribble a few lines of "mathsy" stuff on paper and expect the readers to shower praise on you; your stuff has to stand up to criticism and analysis—it's either right, or wrong.

The myth is also partly the result of the current educational system. Mathematics syllabuses are too full and therefore rely far too heavily on memory. Students know a great deal, but they are given too little time to explore and develop *understanding*. If you know how a given process works, it is surprising how often you can apply the same process in different situations, and when you do, it is very satisfying to know that *you* solved the problem.

1.2   What's It All About?

1.2.1 x?

Next to your signature it's affectionate; it registers a vote; and if you are lucky it shows the location of the treasure. It's one of the easiest letters to write; hence, it was used as a signature. I never did understand that: if anyone could write it, how did you know who did?

But put x in an equation, and it causes nausea and sleeping sickness. This is a shame because that's exactly what it's not supposed to do. So, what the L is an x then? Well, first of all you will notice that we write x like this: "x," not like this: "x," because the latter can easily be mistaken for a multiplication sign. It's a symbol that stands for an unknown quantity of something. We sometimes use y to represent another unknown quantity, so you see we can, and do, use other symbols.

In English we use complicated symbols (words) to represent things: for example, in America we use the word "trunk" to represent the back end of a car. In Britain the word "trunk" refers to the front end of an elephant— among other things! So you see, even symbols which we think we understand are never quite so clear-cut.

So why do we use x? Well, isn't it easier to write

$$x^2 + 3x + 5 = 0$$

rather than

$$\text{trunk}^2 + 3 \times \text{trunk} + 5 = 0$$

or

$$something^2 + 3 \times something + 5 = 0?$$

We use x a lot because it's easy to write, and it saves us a lot of time and effort, *that's all*.

1.2.2   Mathematics?

Mathematics is a language, like English or Spanish or Russian. It allows us to express relationships, not simply brother and sister and parent type relationships, but how things depend on other things. For instance, the area of a rectangular garden can be written

$$area = length \times width$$

except, of course, being mathematicians we'd write

$$a = l \times w$$

because we prefer to write as concisely as possible. In fact, we would probably write $a = lw$ because that's even more economical; we often leave out the multiplication sign when the context is obvious.

Mathematics is also a discipline. It makes us justify everything that we do: What are the assumptions? What can we derive from the assumptions? Can we prove the results? And perhaps more importantly, especially if you are an applied mathematician—which we all are—how can we interpret and use the results?

1.2.3 Functions and Equations

A *function* is a relationship between two or more variables. One of these variables (the *dependent* variable) is defined in terms of the other variable(s) (the *independent* variable(s)). When we know what the relationship is, we can normally write it down as an equation. So, for example, the area of a circle may be represented by the equation

$$A = \pi r^2$$

in which A, the dependent variable, is defined in terms of r, the independent variable.

When we need to calculate the value of the dependent variable, we substitute values for the independent variables and evaluate the resulting equation. This sounds complicated because we are trying to behave like mathematicians. However, consideration of a simple example should make things easier. We can calculate the area of a circle of radius 10 cm by substituting the value of π and the value for r as follows:

$$
\begin{aligned}
A &= 3.14159 \times (10 \text{ cm})^2 \\
&= 314.159 \text{ cm}^2
\end{aligned}
$$

Note that we often omit the units from equations because they can confuse the issue. However, we should always remember to include them in the final answer, and we can use them (or more particularly their dimensions) to advantage when checking equations. (See section 2.3.1, on dimension analysis, below.)

Now you could say "So what?" or "What's the big deal?" and I would reply:

> "Well, I don't need to remember that the area of a circle with a radius of 10 is 314.2, or that the area of a circle of radius 2 is 12.6. Neither do I need a table of such values, nor a carefully plotted calibration curve. All that is necessary is the equation $A = \pi r^2$."

Think of the time, effort, and paper that this function saves! Now think of all the other formulae that you have used and imagine life without them.

One big advantage of mathematics is that you can write something like $A = 3.14159 r^2$ and be understood anywhere in the world. Try writing this relationship in English! Indeed, if we were to send a message into space, in the hope that some alien intelligence will find it and realize that there is intelligence elsewhere, we would stand a much better chance of success if we transmitted the message "$A = 3.14159r^2$" rather than "roses are red." Think about it! The number π has a value which we believe is a universal constant. And while you are in thinking mode, have you noticed that the formula $A = \pi r^2$ is very similar to Einstein's $E = mc^2$?

1.2.4 Relationships

Hopefully you're beginning to get the idea. We use mathematics to express how things relate to each other. As scientists we spend our time looking for these relationships (e.g., $A = \pi r^2$). Isn't it therefore sensible to be able to write them down clearly, so that others can share our discoveries and use them to extend the general knowledge in order to build things (like iPhones) and solve new problems? If you have done the hard work, you may as well document it in the best possible way.

1.2.5 Why Don't We Speak Mathematics All the Time?

Well, while it's good for describing precise scientific things, it's not too good for chatting about everyday things like football and dancing and pop music and stuff. For that we prefer a more evocative language in which we can exercise a bit of imagination.

English can often be interpreted in different ways. For example, the phrase "shed load," which occurs in the context of road traffic reports, makes us smile: "How big a shed?" And a report on the problems of increasing weight brought forth the phrase "the ballooning weight problem," to which a reporter interjected, "What is the optimum weight for ballooning?"

Mathematics isn't like that. When we speak (or more usually write) mathematics, we are trying to be precise and unambiguous. We want to be clearly understood and concise.

1.2.6 And Why Do I Need to Understand It?

A couple of simple examples might just help.

Suppose you go to a shop and buy a package of soap at $3 and two bottles of shampoo at $2 each. You watch the cashier type in $3 + 2 \times 2$, and the register display shows that you owe $10. Do you pay it? First of all, try typing the expression as it stands into your calculator. Depending upon the age and type of your calculator you will produce one of two answers: either $7 or $10. Which is correct? If you know the answer, your mathematics is OK. If you don't, you are likely paying a lot more than you need to! In any event don't you think you owe it to yourself to check these things now and again? (If you don't, please could you send the author a cutting from the tree on which your money grows?)

As another example: suppose that your financial advisor suggests that you take out a bond which will give a 60% return after 10 years rather than invest at 5% per year for 10 years; after all, 10×5 is only 50! What should you do? If you know the answer to this one, your mathematics is very good. But do you understand the calculation? If you do, you may be able to skip a few chapters; otherwise, read on.

1.3 Working with Equations

An equation consists of two expressions separated by an equality sign (=), though sometimes the separator may be an inequality sign (e.g., $\leq$).

1.3.1 Rearranging Equations

There are lots of so-called rules regarding the manipulation of equations. Ignore them! All that you need to remember is that you can do almost anything you like to an equation—provided that you *treat both sides of the equation in the same way.* So that you can add 10 to *both* sides, subtract 23 from *both* sides, or add x to *both* sides. Similarly, you can multiply or divide *both* sides by anything, take the logarithm of *both* sides, or whatever. Just be careful about multiplying or dividing by zero, that's all.

So, for example, if we start out with the equation

$$y = mx + c$$

and would like to find an equation for x, we could subtract c from both sides to give

$$
\begin{aligned}
y - c &= mx + c - c \\
y - c &= mx
\end{aligned}
$$

and then divide both sides by m to give

$$\frac{y - c}{m} = \frac{mx}{m}$$

Then, by cancelling on the right-hand side (RHS)

$$\frac{y - c}{m} = \frac{\cancel{m}\, x}{\cancel{m}}$$

and writing the equation in reverse order we get

$$x = \frac{y - c}{m}$$

The *only* thing to remember is that whatever you do to one side of the equation, you must also do to the other. *But* beware of multiplying or dividing by zero because while $0 \times 2 = 0 \times 3$ is true, the result of dividing both sides by zero $(2 = 3)$ *is not*! The use of this simple trick allows charlatans to accomplish all manner of things!

1.3.2   Order of Evaluating Algebraic Expressions

When we evaluate numerical expressions such as $3 + (2 + 3)^2$ we do not, in general, work from left to right; see the following examples. Mathematical operations are carried out in a strict order as defined by the priority list in table 1.1. Operators high in the list are completed before those lower down. When two operators have the same priority it is customary (but not mandatory) to work from left to right. (In computer programs this may not be the case.)

Table 1.1: Priority of Mathematical Operators

Priority	Operation
1	brackets ( )
2	indices, powers, exponents
3	$\times, \div, /$
4	$+, -$

Be especially careful with expressions like $a \div b \div c$, since they can be easily misinterpreted. Does this expression mean $a \div (bc)$, which results from evaluating $a \div b$ first, or $(ac) \div b$ when $b \div c$ is evaluated first? In cases like this, you should use brackets to make things clear.

But don't develop the habit of using brackets unnecessarily either; a plethora of brackets can be more confusing than none at all.

Examples

1. Evaluate $3 \times (2 + 3)^2$

$$
\begin{aligned}
3 \times (2 + 3)^2 &= 3 \times (5)^2 \\
&= 3 \times 25 \\
&= 75
\end{aligned}
$$

2. Evaluate $3 \times 2 + 3^2$

$$
\begin{aligned}
3 \times 2 + 3^2 &= 3 \times 2 + 9 \\
&= 6 + 9 \\
&= 15
\end{aligned}
$$

1.3.3 Some Useful Algebraic Relationships

Table 1.2: Algebraic Relationships

$x + y = y + x$	Commutative law of addition
$x + (y + z) = (x + y) + z$	Associative law of addition
$x \times y = y \times x$	Commutative law of multiplication
$x \times (y \times z) = (x \times y) \times z$	Associative law of multiplication
$x(y + z) = xy + xz$	
$x + 0 = x$	Definition of zero
$x \times 1 = x$	Definition of unity
$\frac{x+y}{z} = \frac{x}{z} + \frac{y}{z}$	
$\frac{x}{y} \times \frac{v}{w} = \frac{xv}{yw}$	
$\frac{x}{y} + \frac{v}{w} = \frac{xw+yv}{yw}$	

1.3.4 A Word about Calculators

Calculators, like computers, can save a lot of work. Indeed, they allow us to perform tasks which are impossible with only pen and paper. However, as with all things, there are disadvantages. The main one of these is that they take us (the users) a little bit further from understanding the processes of arithmetic. Here are a couple of problems which I have come across when talking mathematics with students.

1. Most modern calculators understand the priorities that govern the order in which calculations are carried out. Older calculators didn't! So if you've inherited your calculator from Mom or Dad, be careful! In the old days *you* were expected to enter the problem in the correct order. If in doubt about your calculator, try entering the following in the order presented:

$$2 + 3 \times 5 =$$

On a modern machine you will get 17 (correctly). On an old-type calculator, you would produce an incorrect result of 25, the reason being that the calculator performs the operations as you enter them:

$$(2 + 3 \Longrightarrow 5) \times 5 \Longrightarrow 25$$

In order to obtain the correct result on such a machine you should type the problem as

$$3 \times 5 + 2 =$$

2. Misunderstanding of the "EXP" key frequently results in errors.

   2 EXP 3 means 2×10^3 *not* 2^3 and will probably appear on your calculator as 2E3. The expression 2^3 (i.e., $2 \times 2 \times 2$) is usually performed using the $\wedge$ or x^y key as follows:

   $$2 \wedge 3 = 8 \qquad \text{or} \qquad 2 \; x^y \; 3 = 8$$

If you have any doubts about the operation of a calculator, try a simple calculation for which you know the answer.

1.4   Exercises

1. If

$$\frac{a}{b} = \frac{c}{d} + e$$

find expressions for each of the five variables in terms of the other four.

2. The equations of motion for a body acted upon by a constant force are

$$
\begin{align}
v &= u + at & (1.1)\\
v^2 &= u^2 + 2as & (1.2)\\
s &= ut + \frac{1}{2}at^2 & (1.3)\\
s &= \left(\frac{u+v}{2}\right)t & (1.4)
\end{align}
$$

where u is the initial velocity, v the final velocity, s is the distance travelled, t is the time, and a the force applied.

   (a) Find an expression for a from equation 1.1.

   (b) Find an expression for u from equation 1.2.

   (c) Find an expression for a from equation 1.3.

   (d) Find an expression for v from equation 1.4.

3. The time in seconds taken by a pendulum to swing back and forth and return to its original position is known as its period, and it may be calculated from the following expression:

$$t = 2\pi \sqrt{\frac{l}{g}}$$

where t is the period (s), l is the length of the pendulum (m) and g is the gravitational constant (9.81 m s^{-2}).

(a) Find an expression for l in terms of t and g, and hence calculate the length of a pendulum that will have a period of one second.

(b) How might we use the formula to calculate the gravitational constant on the moon, should we be fortunate enough to get there?

only in terms of *complex numbers*. Complex numbers are imaginary things which help us to think about the square roots of negative numbers; that's why they are often referred to as *imaginary numbers.*

Real numbers may be represented by points on a straight line:

This line is referred to as the *real number line*. The dashes at each end indicate that it may be extended in either direction indefinitely.

Drawing infinitely long lines or writing indefinitely large numbers is impossible. However, it is sometimes necessary to represent these concepts, and so we have invented the symbol ∞, which represents an unimaginably large number that we call "infinity." We could redraw the real number line using this symbol as follows:

$$-\infty \longleftarrow \overset{\textstyle|}{\underset{-2}{}} \quad \overset{\textstyle|}{\underset{-1}{}} \quad \overset{\textstyle|}{\underset{0}{}} \quad \overset{\textstyle|}{\underset{1}{}} \quad \overset{\textstyle|}{\underset{2}{}} \longrightarrow +\infty$$

We must be careful not to use the symbols $\pm\infty$ to refer to specific values; they are simply symbols of hugeness beyond measure.

2.1   Decimal Number Representation

Some examples of numbers (in decimal notation) are

$$234$$
$$3.14159$$
$$2.\dot{6}$$
$$5000000$$

The dot over the 6 in the third number indicates that the 6 should be repeated indefinitely; an alternative way of writing this would be $2.666\ldots$. An even better and more concise form is $2\frac{2}{3}$. It is interesting to note that this number can never be expressed accurately in decimal; some numbers are like that. Indeed, you should be aware that the vast majority of numbers cannot be represented accurately within a computer. Why not?[1]

We live in an approximate world, a place that mathematicians and physicists refused to believe in for a long time!

---

[1]Because numbers in a computer are stored using a fixed number of significant figures, usually about 7. Therefore, the best that could be done with $2\frac{2}{3}$ would be 2.666667.

2.1.1 Significant Figures and Decimal Places

The second of the numbers in our list above represents π to six significant figures, and to five decimal places, because it contains six digits in total and has been rounded to five digits after the decimal point. The last number has been represented to seven significant figures and no decimal places. The third number is represented to an infinite number of significant figures and decimal places, but we don't normally do that sort of thing, and we certainly cannot on a computer.

2.1.2 Scientific Notation

Accountants would write the last number as 5,000,000 in order to assist the reading. We (scientists) do not, though we often deal with very large or very small values. We get around the problem of writing such numbers by using "scientific notation," in which we represent the number five million as

$$5.0 \times 10^6$$

or sometimes, especially if the number is on a calculator or a computer screen, as $0.5E7$ or $0.5 + 7$ or $0.5_{10}7$. There are several variations on these, depending upon the calculator or computer program in use, but basically the number is represented either by a fraction or a number between 1 and 10 (the *mantissa*, 5.0 in this case) multiplied by a power of 10 (the *exponent*, 6 in this case). Remember that multiplying by 10 shifts the decimal place one digit to the right, multiplying by 10^2 shifts the decimal place two digits to the right, and so on. If the power of 10 is negative, we shift the decimal point the opposite way—i.e., to the left. Here are some numbers expressed in this way:

Distance of the earth from the sun	1.5×10^{11} m
	(150000000000 m)
Diameter of the earth	1.2756×10^7 m
	(12756000 m)
Solar constant	1.366×10^3 Wm^{-2}
	(1366 Wm^{-2})
Mass of the hydrogen atom	1.67×10^{-24} g
	(0.00000000000000000000000167 g)
Speed of light	2.9979×10^8 m s^{-1}
	(299790000m s^{-1})
Avogadro's number	6.02214×10^{23}
	(602214000000000000000000)
Age of the earth	4×10^9 years
	(4000000000 years)
Age of the universe	1.5×10^{10} years
	(15000000000 years)

2.2   Binary and Hexadecimal Numbers

Representation of binary and hexadecimal numbers is fundamental to understanding how computers store data. All modern computers use binary notation in order to represent and operate upon numbers. A brief understanding of binary is therefore helpful in understanding how computers work, while hexadecimals provide a convenient representation of large binary numbers. You may have no interest in the internal workings of a computer, and I can sympathize with that, in which case you could safely skip this section, though you may still find it enlightening to see how numbers can be represented using a different base.

The base, or *radix,* of a number system is the number of different digits, including zero, that the system uses. The decimal system uses 10 different digits, and all numbers are represented as a sequence of powers of 10. As a reminder consider the number 538, which we interpret to mean:

$$5 \times 10 \text{ to the power } 2 \ (\ 5 \times 100 = 500 \ )$$
$$+ \ 3 \times 10 \text{ to the power } 1 \ (\ 3 \times 10 \ \ = \ \ 30 \ )$$
$$+ \ 8 \times 10 \text{ to the power } 0 \ (\ 8 \times 1 \ \ \ = \ \ \ 8 \ )$$

This system has been accepted more by accident than for logical reasons. In fact, there are many arguments in favor of different number systems, but until the advent of the computer there was little pressure to understand any of the alternatives. Why then should we consider them now?

Fundamentally, the major difficulty in computing with decimal numbers is the necessity of providing 10 symbols to allow the representation of numbers. These symbols are of course the familiar digits $0,1, \ldots , 9$. If we were to build a computer based on this system we would need electronic devices that could distinguish between 10 different states. This is difficult! It is much easier (and more reliable) to build electronic devices which recognize only two states: current is either flowing or not, a magnetic field is polarized in one direction or the other, a switch is on or off, and so on. For this reason computers are built using binary logic, in which all information is stored as strings of 1's and 0's, and so if we are to understand the working of a computer, it will be helpful to have some knowledge of binary arithmetic.

2.2.1   Binary Numbers

We are used to representing large numbers as a sequence of powers of 10. For example, 231 means $2 \times 10^2 + 2 \times 10^1 + 1 \times 10^0$. In the binary system we represent numbers as a sequence of decreasing powers of 2, so that the binary number 10011 is interpreted as

$$1 \times 2 \text{ to the power } 4$$
$$+ \ \ \ \ 0 \times 2 \text{ to the power } 3$$

$$+ \quad 0 \times 2 \text{ to the power } 2$$
$$+ \quad 1 \times 2 \text{ to the power } 1$$
$$+ \quad 1 \times 2 \text{ to the power } 0$$

and is the equivalent of the decimal number 19.

It is obvious that, though we need only two symbols (0 and 1) in binary, in general the binary representation of a number will require more digits than the equivalent decimal representation. A few examples may be helpful:

Decimal	Binary
5	101
0	0
65	1000001
31	11111
10.5	1010.1
3.25	11.01

Within binary numbers the period is known as the *binary point*.

2.2.2 Conversion from Decimal to Binary

In order to convert the decimal number 13 to binary we repeatedly divide by 2 as follows:

$$13/2 = 6 \quad \text{remainder } 1$$
$$6/2 = 3 \quad \text{remainder } 0$$
$$3/2 = 1 \quad \text{remainder } 1$$
$$1/2 = 0 \quad \text{remainder } 1$$

Now if we read the remainder column starting from the bottom, we have 1101, which is the binary equivalent of the decimal number 13.

2.2.3 Hexadecimal Numbers

In the hexadecimal system we use the number 16 as the base (or radix), as opposed to 10 in the normal decimal system or 2 in the binary system. This means that we have to "invent" six new symbols to represent the additional digits. The hexadecimal digits are as follows:

$$0, 1, 2, \ldots, 9, A, B, C, D, E, F$$

so that

$$A_{16} = 10_{10}$$
$$B_{16} = 11_{10}$$
$$\vdots \qquad \vdots$$
$$F_{16} = 15_{10}$$

where the subscript indicates the value of the radix.

2.2.4   Conversion from Decimal to Hexadecimal

The process of converting decimal to hexadecimal is similar to that of converting decimal to binary:

$$123_{10}/16_{10} \; = \; 7_{10} \quad \text{remainder } 11_{10} = B_{16}$$
$$7_{10}/16_{10} \; = \; 0_{10} \quad \text{remainder } 7_{10} \; = 7_{16}$$

Reading the remainder column from the bottom gives the converted value:

$$123_{10} = 7B_{16}$$

2.2.5   Binary-Hex Conversion

Each hexadecimal digit may be represented by four binary digits, since $2^4 = 16$. A table of hexadecimal and their equivalent binary representations is given below:

Hexadecimal	Binary
0	0000
1	0001
2	0010
3	0011
4	0100
5	0101
6	0110
7	0111
8	1000
9	1001
A	1010
B	1011
C	1100
D	1101
E	1110
F	1111

The convenience of hexadecimal numbers is that we can translate easily between "hex" and binary, simply by replacing each hex digit with its corresponding (four-digit) binary equivalent. Thus, if we want to examine a number held within a computer, we can easily expand the hex to give the individual binary digits, yet we can print the values in hex to save space.

The basic unit in terms of electronic data transfer is the *byte*, which consists of eight binary bits or two hex characters. A byte can therefore be expressed as a two-digit hex number.

2.3 Preliminary Calculations: Check the Problem

It is very tempting to rush into a computation with calculator in hand. However, it is often useful (and safer) to perform some simple arithmetic on the back of an envelope first.

There have been many cases where a quick check of the calculations would have avoided serious consequences. It is very easy to make mistakes when keying in numbers, or to misinterpret results from a calculator or spreadsheet. When such calculations may have serious consequences, you should check. Miscalculating a drug dose by a factor of 10 is a serious (and real) example.

There is also, of course, the possibility that your computer or calculator may produce the wrong answer! How dare anyone even think such a thing?!

I will illustrate what can be done by looking at an example from geo-chemistry. Geochemists use a relationship called *residence time* in order to elicit information about the movement of chemical elements through large systems like the oceans.

$$\text{residence time} = \frac{\text{mass of element in oceans}}{\text{mass turnover per year}} \tag{2.1}$$

$$= \frac{\text{conc'n in oceans} \times \text{volume of oceans}}{\text{conc'n in rivers} \times \text{flux from rivers}} \tag{2.2}$$

Residence time is the length of time that an element remains in the ocean.

2.3.1 Dimension Analysis

A simple check which we can perform on formulae like equation (2.1) above is to examine the dimensions involved. Note here that we are talking about *dimensions* (e.g., mass, length, time, electric current), not *units*. In order to carry out this check we replace the terms in the formula by their dimensions, mass (M), length (L), and time (T).

$$\text{residence time (T)} = \frac{\text{mass of element in oceans (M)}}{\text{mass turnover per year (MT}^{-1})}$$

In this case, after cancelling the M's on the RHS, the dimensions on both sides of the equation agree (T), so that we can be reasonably confident that the formula will generate a sensible answer. When the dimensions do not agree, the formula is in error and should be checked.

Application: How Much Rain Flows into the Oceans?

One question that a geochemist might like to ask is "What is the annual flux of rainwater into the oceans?" This could be determined using the quantities and concentrations of calcium (Ca) listed below. If you have trouble with numbers like 0.818×10^6, a quick reference to section 2.1.2 will help.

Residence time of Ca in the oceans 0.818×10^6 years
Concentration of Ca in seawater   412 mg dm^{-3}
Concentration of Ca in river water   15 mg dm^{-3}
Volume of the oceans 1.37×10^{21} dm^3

First we must rearrange equation (2.2) as follows:

$$\text{flux from rivers} \quad = \quad \frac{\text{conc'n of Ca in oceans} \times \text{volume of oceans}}{\text{conc'n of Ca in rivers} \times \text{residence time}}$$

$$= \quad \frac{412 \times 1.37 \times 10^{21}}{15 \times 0.818 \times 10^6}$$

2.3.2   A Rough Calculation on the Back of an Envelope

Evaluating the above expression will have most of us diving for a calculator
or spreadsheet. However, it is a good idea to perform a rough evaluation on
a scrap of paper first, since it is easy to make a mistake entering the values.
We do this by approximating the individual values to the extent that we
can perform a rough calculation using head and pencil only.

You will just have to imagine that the following is a scruffy envelope!

$$\text{flux from rivers} \quad \approx \quad \frac{(4 \times 10^2) \times 1.4 \times 10^{21}}{15 \times 0.8 \times 10^6}$$

$$= \quad \frac{4 \times 1.4}{15 \times 0.8} \times \frac{10^2 \times 10^{21}}{10^6}$$

$$= \quad \frac{5.6}{12} \times 10^{17}$$

$$\approx \quad \frac{1}{2} \times 10^{17} \text{ dm}^3 \text{ year}^{-1}$$

Performing the calculation correctly on a calculator produces the result
4.6×10^{16} (or 0.46×10^{17}). The agreement between this and the rough answer
should make you feel confident that it is correct.

2.4 Exercises

1. Represent the following numbers in scientific notation with a mantissa in the range 1 to 9.$\dot{9}$.

   (a) 12.63
   (b) 8000000
   (c) $10\frac{1}{2}$
   (d) $3\frac{1}{3}$
   (e) 1/100

2. Complete the following table of binary, decimal, and hexadecimal numbers.

Binary	Decimal	Hex
1011		
	13	
		42A

3. What is the result of dividing the hexadecimal number $17DDE$ by 2?

2.5   Answers

1. (a) 1.263×10^1 or $1.263E1$ or $1.263_{10}1$

   (b) 8×10^6

   (c) 1.05×10^1

   (d) 3.333333×10^0

   (e) $1 \times 10^{-}2$

2. The completed table is

Binary	Decimal	Hex
1011	11	B
1101	13	D
010000101010	1066	$42A$

3. The result of dividing the hexadecimal number $17DDE$ by 2 is BEEF.

Chapter 3

Powers and Logarithms

The main reason for this chapter is to show you that difficult formulae (which you have been taught to remember) are not thought up in isolation. They are usually derived from a few simple assumptions, from which they follow logically. Once you have followed the logic, life becomes easier because you understand why the formulae work. Not only that; if pushed you can derive them again yourself without having to fill your head with unnecessary details. Memory is no substitute for understanding; it helps—but you can always look things up if you know where to search.

Your parents and grandparents hated logarithms ("logs") because performing calculations using log tables was difficult and tedious. Now for the good news: no one of sound mind does this any more! However, the manipulation of expressions involving powers is an essential skill if you are to deal with scientific calculations. Once you have gained that skill, understanding logarithms follows naturally.

Now read on: understanding powers and logarithms is good, but doing arithmetic with logs is a waste of time; calculators do it better!

3.1   Powers and Indices

We define a^2 as the number resulting when two copies of a are multiplied together:

$$a^2 = a \times a$$

Similarly,

$$a^3 = a \times a \times a$$

Now for the difficult bit. We could go on defining $a^4, a^5, \ldots$, but this is a pain, so we will think about the general case, $a^{anything}$. We define a^m, where m is any number, as the number resulting when m copies of a are all multiplied together:

$$a^m = a \times a \times a \times \cdots \times a$$

The term m is called the *power* or *index*, and a is referred to as the *base*. In an equation like the one above where several things are to be multiplied together, the individual terms are referred to as *factors*. We refer to the expression a^m as a to the power m, or "a to the m" for short. Expressions like a^m, 3^2, and 10^{-13} are called *exponential expressions*.

3.1.1   Some General Rules of Powers and Indices

We shall restrict ourselves to the case where a is any real number other than zero and m and n are integers, though the following results can be shown to apply more generally.

$a^m \times a^n$

$$a^m = \underbrace{a \times a \times \cdots \times a}_{m \text{ factors}}$$

$$a^n = \underbrace{a \times a \times \cdots \times a}_{n \text{ factors}}$$

$$\therefore a^m \times a^n = \underbrace{a \times a \times \cdots \times a}_{m+n \text{ factors}}$$

$$\therefore a^m \times a^n = a^{m+n}$$

$$\text{e.g.,} \qquad a^2 \times a^3 = a^{2+3}$$
$$= a^5$$

$$\text{e.g.,} \qquad 10^5 \times 10^2 = 10^7$$

a^0

$$a^m \times a^0 = a^{m+0} = a^m$$

$$\therefore a^0 = 1$$

because multiplying by a^0 leaves any value unchanged.

a^{-m}

$$a^m \times a^{-m} = a^{m-m} = a^0 = 1$$

$$\therefore a^{-m} = \frac{1}{a^m}$$

a^m/a^n

$$\frac{a^m}{a^n} = \frac{a \times a \times \cdots a}{a \times a \times \cdots a} \qquad \frac{m \text{ factors}}{n \text{ factors}}$$

$$= a \times a \times \cdots \times a \qquad (m - n) \text{ factors}$$

$$\therefore \frac{a^m}{a^n} = a^{m-n}$$

$(a^m)^n$

$$(a^m)^n = a^m \times a^m \times \cdots \times a^m \quad (n \text{ factors})$$

$$= a^{m+m+\cdots+m}$$

$$\therefore (a^m)^n = a^{m \times n}$$

e.g., $\qquad (10^2)^3 = 10^6$

$a^{1/m}$

$$\text{Consider } a^{1/m} \times a^{1/m} \times \cdots \times a^{1/m} \qquad (m \text{ factors})$$

$$= a^{(1/m+1/m+\cdots+1/m)}$$

$$= a^1$$

$$(a^{1/m})^m = a$$

$$\therefore a^{1/m} = \sqrt[m]{a}$$

3.1.2 Rules of Powers and Indices: Summary

$$a^m = a \times a \times a \times \cdots \times a$$

$$a^m \times a^n = a^{m+n}$$

$$a^0 = 1$$

$$a^{-m} = \frac{1}{a^m}$$

$$\frac{a^m}{a^n} = a^{m-n}$$

$$(a^m)^n = a^{m \times n}$$

$$a^{1/m} = \sqrt[m]{a}$$

3.2   Logarithms

In the "olden days" all arithmetic was done using pencil, paper, and the stuff which keeps your ears apart. This was hard, and therefore any tool or magic spell which could facilitate arithmetic was perceived to be a very good thing.

Logarithms were invented by a Scotsman named John Napier, who probably had nothing better to do during the long winter nights than chuck the odd log under his still. He realized that by representing numbers as powers of 10 and then using the rules derived in the last section, he could avoid a lot of tedious arithmetic (idle mathematicians again!).

His idea was to produce a look-up table relating x and p where $x = 10^p$ (log tables) and a reverse or antilog table which would allow the user to find x given p. Then, in order to calculate $x \times y$, one would

1. look up the logarithm of x (p, say) and the logarithm of y (q);

2. calculate $r = p + q$;

3. look up the antilog of r to give z, where $z = x \times y$.

So, for example, to calculate 45.67×23.45,

1. look up $\log 45.67$ ($= 1.6596$) and $\log 23.45$ ($= 1.3701$);

2. add them together to give 3.0297;

3. look up the antilog of 3.0297, which gives the answer 1071 to four significant digits.

Thus, long multiplication was reduced to looking up three values in tables, together with one addition. This saved a lot of work and reduced the chances of error. Further exploitation of the properties of exponentials allow the simplification of other arithmetic calculations.

This was important because rapid calculations, together with accurate timing, were essential to navigation. Good navigation, in turn, was the key to global power. (Actually, it meant that entrepreneurs could pilfer all manner of treasure throughout the world and then find their way back home!)

Until the 1970s every science student had a book of log tables, though some people, usually engineers, had fancy gadgets called slide rules, which were a sort of automated set of log tables. Mathematicians and "real scientists" referred to slide rules as "guessing sticks." The equivalent of log and antilog tables are the "log" and "10^x" buttons on your calculator. Nowadays log tables that have survived fire-lighting are confined to antiquarian bookshops. Electronic pocket calculators still use logarithms but do so largely in secret. Sometimes you may notice strange numbers appear while you are performing calculations; these are probably logarithms used by your calculator during the intermediate steps of a calculation.

To a large extent the overt use of logarithms can now be avoided because of the power of the computer. However, there are still many applications and protocols which make use of the old-fashioned ways, and so it is both useful and informative to have a basic understanding of how logarithms work.

3.2.1 What Are Logarithms?

We shall start by discussing logarithms to base 10 because they are widely used. However, it should be borne in mind that we could use logs to any base, though the only other base of any consequence is that of e, the exponential constant $2.718282\ldots$

Basically $\log(x)$ is the number of 10s that you have to multiply together to generate the value x. Thus,

$$\log(100) = 2 \quad \text{because} \quad 100 = 10 \times 10$$

In the same way

$$\log(1000000) = 6 \quad \text{because} \quad 1000000 = 10 \times 10 \times 10 \times 10 \times 10 \times 10$$

So far this is easy. But you (and I) have difficulty understanding $\log(50)$. It is bigger than 1 because $10^1 = 10$ and is less than 2 because $10^2 = 100$. (Actually, it is 1.6990 to four decimal places.) Incidentally, if you calculate $\log(500)$ you will notice that it is exactly 1 larger than $\log(50)$—2.6990. This is because you need to multiply by one more 10 to get 500 than you do to get 50 ($500 = 50 \times 10$). Likewise, $\log(5000)$ is 1 more than $\log(500)$. Notice that numbers formed by repeatedly multiplying by 10 have logarithms which increase by 1 at each multiplication.

Repeated multiplication by any number produces a sequence of values whose logarithms behave in the same way—that is, their log values increase by a constant amount at each multiplication. Consider the sequence 5, 25, 125, ..., and the corresponding log values:

x	5	25	125	625
$\log(x)$	0.6990	1.3979	2.0969	2.7959

Notice that as the number is multiplied at each step by 5, the corresponding logarithm is increased by 0.6990, because $\log(5) = 0.6990$.

Many physical relationships involve repeated multiplication: growth of a population over successive generations, compound interest, radioactive decay, repeated dilutions, and the frequencies of successive notes in the music scale. Plotting a simple graph of the logarithm of these values against time should reveal a straight-line relationship because the logarithm will increase (or decrease) at a constant rate.

3.2.2   Definition

The value of a logarithm depends upon two things: the number itself and the *base* of the logarithms.

The logarithm of N to the base a is usually written $\log_a N$ and is defined as follows:

$$\text{if} \quad a^x = N \quad \text{then} \quad \log_a N = x \tag{3.1}$$

It is the power to which the base has to be raised in order to generate the number (i.e., how many a's we need to multiply together to produce the value N). So that

$$N = a^{\log_a N}$$

For example, the logarithm to base 3 of 9 ($\log_3 9$) is 2 because $9 = 3^2$.

$$\text{Examples:} \quad \begin{aligned} 2^4 &= 16 & \therefore \ \log_2 16 = 4 \\ 12^2 &= 144 & \therefore \ \log_{12} 144 = 2 \end{aligned}$$

Common Logarithms

Traditionally, tables of logarithms to the base 10 were used, and these were known as *common logarithms*.

Natural Logarithms

Mathematicians and physicists tend to use logarithms to the base e; e is the number 2.718282 ..., which is sometimes referred to as *Euler's number* or the *exponential constant*. This may seem strange, but the number e occurs naturally in many areas of science. Logarithms to base e are referred to as *natural logarithms*. The modern equivalent of natural log and antilog tables are the ln and e^x buttons on your calculator.

The exponential function $\exp(x)$ returns the value of e^x so that the two expressions are equivalent.

Logarithms to base 10 and e occur so often that they are abbreviated as follows:

$$\begin{aligned} \log(x) &\equiv \log_{10}(x) \\ \ln(x) &\equiv \log_e(x) \end{aligned}$$

Generally you are advised to work with natural logarithms because the algebra is easier, though sometimes (e.g., when calculating pH) you will find it more convenient to work to base 10.

e: An Interesting Number

Suppose that your bank suggested an investment for a specified period, at the end of which your investment would be returned together with 100% interest. Thus, if you invested $1, at the end of the period you would collect $2.

If you were able to persuade the bank to pay you at half the rate, but compounded over two half periods, your final return would be

$$(1 + 1/2) * (1 + 1/2) = (1 + 1/2)^2 = \$2.25$$

since you would receive $0.50 interest halfway through the period, and then you would have $1.50 invested over the remaining half period.

You might even persuade the bank to give you interest every quarter, in which case you would receive

$$(1 + 1/4) * (1 + 1/4) * (1 + 1/4) * (1 + 1/4) = (1 + 1/4)^4 = \$2.44$$

Being a thinking person you would obviously see advantage in taking this further. A general formula when the period is split into n equal intervals is

$$(1 + 1/n)^n$$

from which you could calculate the final values shown below.

n	Final value
1	2.00
2	2.25
4	2.44
10	2.59
100	2.70

It is clear that as the value of n increases, the result of evaluating $(1 + 1/n)^n$ approaches some limiting value. As mathematicians we write this as follows:

$$\lim_{n \to \infty} (1 + 1/n)^n = 2.718281828\ldots$$

by which we mean that the limiting value of this expression, if we were able to calculate it with n taking an infinite value (which we can't), would be $2.718281828\ldots$ We represent this limiting value as e and refer to it as Euler's number or simply e.

3.2.3 Mathematical Derivation of the Rules of Logarithms

In order to demonstrate that the following rules are independent of the base chosen we shall work with logarithms to the base a, where a is any positive number. These rules are closely related to those for *powers*.

$\log_a 1$

$$\log_a 1 = 0 \quad \text{because} \quad a^0 = 1$$
$$\ln(1) = 0 \quad \text{because} \quad e^0 = 1$$
$$\log(1) = 0 \quad \text{because} \quad 10^0 = 1$$

$\log_a a$

$$\log_a a = 1 \quad \text{since} \quad a^1 = a$$

$\log_a(M \times N)$

$$\text{Let} \quad x = \log_a M \quad \therefore a^x = M$$
$$\text{and} \quad y = \log_a N \quad \therefore a^y = N$$

$$M \times N = a^x \times a^y = a^{x+y}$$
$$\log_a(M \times N) = x + y$$
$$\log_a(M \times N) = \log_a M + \log_a N$$

$\log(M/N)$

$$\text{Let} \quad x = \log_a M \quad \therefore a^x = M$$
$$\text{and} \quad y = \log_a N \quad \therefore a^y = N$$

$$\frac{M}{N} = \frac{a^x}{a^y} = a^{x-y}$$

$$\log_a(M/N) = x - y$$
$$\log_a(M/N) = \log_a M - \log_a N$$

$\log_a(M^p)$

$$\text{let} \quad x = \log_a M$$
$$\therefore M = a^x$$
$$\therefore M^p = a^{px}$$
$$\therefore \log_a(M^p) = px$$
$$\log_a(M^p) = p \log_a M$$

Sample Calculations Involving Logarithms

1. Calculate $\log_{10}(10 \times 100)$.

$$\log_{10}(10 \times 100) = \log_{10} 10 + \log_{10} 100 = 1 + 2 = 3$$

2. Calculate $\log_2(2 \times 16)$.

$$
\begin{aligned}
\log_2(2 \times 16) &= \log_2 2 + \log_2 16 \\
&= 1.0 + 4.0 \\
&= 5 \qquad (= \log_2(32))
\end{aligned}
$$

3. Calculate $\log_4(16^3)$.

$$
\begin{aligned}
\log_4(16^3) &= 3 \times \log_4 16 \\
&= 6 \qquad (= \log_4 4096)
\end{aligned}
$$

4. Find the logarithm of $32\sqrt[5]{4}$ to base $2\sqrt{2}$.

Let x be the required logarithm;

then
$$
\begin{aligned}
(2\sqrt{2})^x &= 32\sqrt[5]{4} \\
(2.2^{\frac{1}{2}})^x &= 2^5.2^{\frac{2}{5}} \\
2^{\frac{3}{2}x} &= 2^{5+\frac{2}{5}} \\
\frac{3}{2}x &= \frac{27}{5} \\
x &= \frac{18}{5} = 3.6 \\
\log_{2\sqrt{2}}(32\sqrt[5]{4}) &= 3.6
\end{aligned}
$$

3.2.4 Calculating Logarithms to a Different Base

Suppose we require $\log_b N$ having been given $\log_a N$.

$$
\begin{aligned}
\text{Let } \log_b N &= y \text{ so that } b^y = N \\
\therefore \quad \log_a N &= \log_a(b^y) \\
\therefore \quad \log_a N &= y \log_a b \\
\therefore \quad y &= \frac{\log_a N}{\log_a b} \\
\log_b N &= \frac{\log_a N}{\log_a b} \qquad (3.2)
\end{aligned}
$$

Example: Suppose We Know $\ln(2)$ But Need $\log(2)$.

$$
\log_{10} 2 = \frac{\log_e 2}{\log_e 10} = \frac{0.6931}{2.3026} = 0.3010
$$

3.2.5   Rules of Logarithms: Summary

The following relationships apply, whatever the base (a) of the logarithms.

$$\log_a 1 = 0$$

$$\log_a a = 1$$

$$\log_b X = \frac{\log_a X}{\log_a b}$$

$$\log_a XY = \log_a X + \log_a Y$$

$$\log_a \tfrac{X}{Y} = \log_a X - \log_a Y$$

$$\log_a X^n = n \log_a X$$

3.3   Population Dynamics and the Exponential Equation

The usual basis for the description of population growth is the exponential equation

$$N_t = N_0 \, e^{rt}$$

in which N_0 is the number of individuals in the initial population and r is known as the "intrinsic rate of natural increase of the population." This form of the equation does not lend itself to understanding the basics of population growth. The difficulty is due mainly to the inclusion of the exponential constant e, which is a strange number to those other than mathematicians.

However, if we use the relationship

$$x^{pq} = (x^p)^q$$

we can see that the above expression may be represented as follows:

$$N_t = N_0 R^t$$

where

$$R = e^r$$

In particular, if we now refer to R as the intrinsic rate of increase, R has a much more intuitive meaning, since when R is greater than 1, the population is increasing, while when it is less than 1, the population will decrease. Furthermore, when $R = 1$ the population is stable.

This expression also allows us to see that as t increases by a unit value, the population number will increase by the factor R:

$$N_{t+1} = N_t R$$

since

$$N_0 R^{t+1} = N_0 R^t R$$

3.4 Exercises

1. Evaluate the following expressions using powers of thought and pen or pencil only.

   (a) $16^{\frac{1}{2}}$

   (b) $27^{\frac{2}{3}}$

   (c) $\log 1$

   (d) $\ln e^7$

   (e) $\log_9 3$

   (f) $\log \frac{1}{10000}$

   (g) $\log_2(\frac{1}{8})$

   (h) $\log_q q^5$

   (i) $\log_2 16$

2. Prove: $\log_a b \times \log_b c \times \log_c a = 1$

3. If $6^{3x} = 14.7$, calculate x.

4. The pH of a chemical solution is defined as follows:

   $$pH = -\log_{10}(\text{hydrogen ion concentration})$$

   What is the hydrogen ion concentration of a solution with pH $= 4.2$?

5. Measurements of the concentration C of a substance are to be taken at several values of time t after the start of an experiment. It is expected that the values of C can be predicted by the equation

   $$C = C_0 e^{-kt},$$

   where C_0 is the initial concentration, t is the time, and k is a constant whose value is sought. Estimate the value of k from the following data by plotting $\ln C$ against t:

t	0	5	10	15	20
C	20	12	7.3	4.5	2.7

3.5 Answers

1. (a) 4

    (b) 9

    (c) 0

    (d) 7

    (e) $\frac{1}{2}$

    (f) -4

    (g) -3

    (h) 5

    (i) 4

2.

$$
\begin{aligned}
&\log_a b \times \log_b c \times \log_c a \\
= \;&\log_a b \times \frac{\log_a c}{\log_a b} \times \frac{\log_a a}{\log_a c} \\
= \;&\log_a a \\
= \;&1
\end{aligned}
$$

3.

$$
\begin{aligned}
6^{3x} &= 14.7 \\
\therefore\; 3x \log 6 &= \log 14.7 \\
x &= \frac{\log 14.7}{\log 6} \times \frac{1}{3}
\end{aligned}
$$

4. Let H^+ be the hydrogen ion concentration.

$$
\begin{aligned}
4.2 &= -\log H^+ \\
-4.2 &= \log H^+ \\
H^+ &= 10^{-4.2} \\
H^+ &= 6.310 \times 10^{-5}
\end{aligned}
$$

5. Taking logarithms of both sides of the equation we have

$$
\ln C = \ln C_0 - kt
$$

Plotting $\ln C$ against t, we obtain a straight line whose slope is -0.1. Hence $k = 0.1$

Chapter 4

Calculations and Applications

My aim in this chapter is to convince you that the mathematics that you have learned so far is useful and can help solve problems in a wide range of applications. I hope that somewhere in what follows you will find at least some examples that are relevant to you and that you will follow the calculations through, in order to gain confidence in the knowledge that you have gained. Remember that the objective is to understand the mathematics and its application.

4.1 Convert Miles/Hour (mph or miles hour^{-1}) to m s^{-1}

The way to tackle unit conversion problems is to work through, changing one unit at a time, in successive stages as follows:

$$\frac{\text{miles}}{\text{hour}} \times 1.609 \Rightarrow \frac{\text{km}}{\text{hour}} \times 1000 \Rightarrow \frac{\text{m}}{\text{hour}} \times \frac{1}{60 \times 60} \Rightarrow \frac{\text{m}}{\text{s}}$$

Combining all these factors produces the formula

$$\frac{\text{miles}}{\text{hour}} \times \frac{1.609 \times 1000}{60 \times 60} \Rightarrow \frac{\text{m}}{\text{s}}$$

A "back of the envelope" calculation gives

$$\frac{1.609 \times 1000}{60 \times 60} = \frac{1.609 \times 10}{6 \times 6} = \frac{16.09}{36} \approx 0.5$$

which the calculator confirms to give

$$\frac{\text{miles}}{\text{hour}} \times 0.447 \Rightarrow \frac{\text{m}}{\text{s}}$$

Hence, to convert a speed in miles/hour to m s^{-1}, we multiply by 0.447 (to three significant figures).

4.2 Body Mass Index (lb/in^2)

The body mass index is defined in Europe as the ratio of mass (kg) divided by height squared (m^2). For some of us the units are inconvenient and we may prefer to perform the calculation in imperial units. The conversion factor can be found as follows:

$$\begin{aligned}
\text{BMI} &= \frac{\text{weight(kg)}}{\text{height}^2(\text{m}^2)} \\[2mm]
&= \frac{\text{weight(lb)} \times 0.4536}{\text{height}^2(\text{in})^2 \times (2.54 \times 10^{-2})^2} \\[2mm]
&= \frac{\text{weight(lb)} \times 0.4536}{\text{height(in)}^2 \times 6.452 \times 10^{-4}} \\[2mm]
&= \frac{\text{weight(lb)}}{\text{height(in)}^2} \times 703
\end{aligned}$$

4.3 The pH of a Solution

The pH of a solution is the negative of the logarithm to base 10 of the hydrogen ion activity. If a biological fluid has a pH of 7.5, what is the hydrogen ion activity?

$$\begin{aligned}
7.5 &= -\log_{10}(H^+) \\
\therefore H^+ &= 10^{-7.5} \quad \text{from the definition of logarithm [eq. (3.1)]} \\
&= 3.162 \times 10^{-8}
\end{aligned}$$

4.4 How Many Microbes? The Viable Count Method

Microbiologists spend much of their time identifying, culturing, and counting microbes. The presence and concentration of microbes will determine whether action needs to be taken to prevent food spoilage, infection, etc.

In the viable count method different concentrations of microbes are made up by successively diluting a sample solution by a constant factor $1/f$, usually $f = 10$. A small sample from each dilution is plated and the plates incubated. Counts are made of the cfu's (colony-forming units) present in each of the dilutions after a period of culture. The plate containing between 30 and 300 cfu's is normally regarded as the sample to be used in calculating the original concentration. If, for this sample,

$$\begin{aligned}
n &= \text{number of cfu's counted} \\
d &= \text{number of dilutions for this plate} \\
v &= \text{volume of sample plated (ml)} \\
f > 1 &= \text{the dilution factor}
\end{aligned}$$

then the concentration, c, of cells in the original sample can be calculated as follows:

$$c = \left(\frac{n}{v}\right) \times f^d \text{ ml}^{-1}$$

The term n/v gives the concentration (ml^{-1}) of cfu's in the plated sample, while the term f^d scales the count according to the number of dilutions.

Thus, if we counted 40 cfu's on a plate that had five dilutions of factor 10, and the plate sample volume was 0.05 ml, the calculation would be

$$c = (40/.05) \times 10^5 = 8 \times 10^7 \text{ ml}^{-1}$$

Note that 50 $\mu l = 0.05$ ml.

4.5 Surface Area of Humans

Dubois and Dubois[1] showed that the surface area of the human body may be approximated by the formula

$$S = 0.007184 W^{0.425} H^{0.725} \tag{4.1}$$

where $S =$ surface area (m^2), $W =$ weight (kg), and $H =$ height (cm). Thus, the surface area of a person 170 cm tall and weighing 70 kg is predicted to be 1.810 m^2.

Many older people (e.g., professors) and Americans would find this formula extremely difficult. They would prefer to work in terms of feet, pounds and inches, so we may be asked (told!) to provide a simpler version.

If we are to use equation (4.1) with inputs in pounds and inches, we must first of all convert these values into the units expected by the formula, so that if w is the weight in pounds and h the height in inches we have

$$\begin{aligned} W &= \frac{w}{2.2046} \\ H &= 2.540 \times h \end{aligned}$$

since 1 kg $\approx$ 2.2046 lb and 1 in. $\approx$ 2.540 cm. Substitution in equation (4.1) gives

$$S = 0.007184 \left(\frac{w}{2.2046}\right)^{0.425} (2.540 \times h)^{0.725} \tag{4.2}$$

where the input values w and h are now in the required units, but the output is still calculated in terms of square meters. In order to calculate s (the area in square feet) we must introduce the additional conversion

$$S = \frac{s}{3.2808^2}$$

---

[1]D. Dubois and E. F. Dubois, A formula to estimate the approximate surface area if height and weight be known, *Archives of Internal Medicine* 17 (1916): 863-71.

since 1 m $\approx$ 3.2808 ft, and hence, 1 m^2 is 3.2808×3.2808 ft^2.

Substituting this in equation (4.2) gives

$$\frac{s}{3.2808^2} = 0.007184 \left(\frac{w}{2.2046}\right)^{0.425} (2.540 \times h)^{0.725} \qquad (4.3)$$

where all the values are now in the required units. This is extremely messy, so here is the tidying-up in stages:

$$\begin{aligned} s &= 3.2808^2 \times 0.007184 \times \left(\frac{w}{2.2046}\right)^{0.425} \times (2.540h)^{0.725} \\ &= 3.2808^2 \times 0.007184 \times \left(\frac{1}{2.2046^{0.425}} \times w^{0.425}\right) \times (2.540^{0.725} \times h^{0.725}) \\ &= \left[\frac{3.2808^2 \times 0.007184 \times 2.540^{0.725}}{2.2046^{0.425}}\right] \times w^{0.425} \times h^{0.725} \end{aligned}$$

Evaluating the arithmetic gives the required working version of equation (4.1):

$$s = 0.1086 w^{0.425} h^{0.725}$$

Notice that the powers of the weight and height terms remain unaltered; only the initial constant has changed.

4.6 Blood Flow in the Arteries

The velocity of flow, or volume flux, of a fluid along a cylindrical vessel of length l and radius a is predicted by Poiseuille's formula:

$$V = \frac{\pi(\Delta_p)a^4}{8\eta l} \,,$$

where η is a constant characteristic of the fluid (viscosity) and Δ_p is the pressure difference between the ends of the pipe.

This equation has applications in the study of blood flow, food processing, coolant, lubricants, and other fields.

Suppose we reduce the radius of the tube to half its original value; what effect does this have on V?

$$\begin{aligned} V^* &= \frac{\pi(\Delta_p)(a/2)^4}{8\eta l} \\ &= \frac{\pi(\Delta_p)a^4}{8\eta l . 2^4} \\ &= \frac{1}{16} V \end{aligned}$$

The reduction of the radius to one-half reduces the volume flux by a factor of 16. One consequence of this is that your heart must work 16 times harder in order to deliver the same amount of blood if your arteries are reduced in diameter by one-half!

4.7 The Growth of a Bacterial Population

If we assume that a bacterial population is not restricted by space, lack of nutrients, or other factors, the population will double at each generation. Thus, if we have N_0 bacteria at time 0, there will be $N_0 \times 2$ bacteria after one generation period:

$$N_1 = N_0 \times 2$$

Similarly, during the next generation period,

$$
\begin{aligned}
N_2 &= N_1 \times 2 \\
&= N_0 \times 2 \times 2 \\
&= N_0 \times 2^2
\end{aligned}
$$

After t such time intervals the number of bacteria in the population is given by

$$N_t = N_0 \times 2^t$$

For *E. coli* the generation time is 20 min, so that after 20 generations (6 hr 40 min) we can predict the population as follows:

$$
\begin{aligned}
N_{20} &= N_0 \times 2^{20} \\
&= N_0 \times 2^{10} \times 2^{10} \\
&\qquad\qquad (2^{10} = 1024) \\
&\simeq N_0 \times 1 \text{ million}
\end{aligned}
$$

4.8 Light Passing Through a Liquid

The Beer-Lambert law states that when light passes through a solution, the intensity of the emergent light (I) is less than that of the incident light (I_0). This relationship has important applications in spectrophotometry for chemical analysis and is given by

$$I = I_0 10^{-\varepsilon cd}$$

where c is the concentration of the solution (moles/liter), d is the length of the light path through the liquid, and ε is the extinction coefficient.

This equation allows us to measure concentrations using colorimetry as follows:

1. Measure the intensity, I_s, for a standard solution of concentration, c_s, to give

$$I_s = I_0 \, 10^{-\varepsilon c_s d} \tag{4.4}$$

2. Measure the intensity, I_t, of the test solution whose concentration, c_t, is unknown to give

$$I_t = I_0 10^{-\varepsilon c_t d} \tag{4.5}$$

3. Take logs (to base 10) of equations (4.4) and (4.5) and rearrange each equation to give

$$
\begin{aligned}
c_s &= \frac{\log I_0 - \log I_s}{\varepsilon d} \\
c_t &= \frac{\log I_0 - \log I_t}{\varepsilon d} \\
\therefore \frac{c_t}{c_s} &= \frac{\log I_0 - \log I_t}{\log I_0 - \log I_s} \\
\therefore c_t &= c_s \frac{\log(I_0/I_t)}{\log(I_0/I_s)}
\end{aligned}
\tag{4.6}
$$

4.9 A Water Pollution Incident

An organization has been identified as having polluted a lake by dumping a quantity of arsenic into it! The river authority knows exactly when the dump took place and has measured the concentration as 4.00 ppb 10 hr after the incident and 2.80 ppb after 13 hr.

The organization claims that the initial concentration was only 7 ppb, which is below the legal limit of 10 ppb. Do we believe this claim?

We assume that the concentration can be described by an exponential decay relationship as follows:

$$C = C_0 e^{-kt}$$

where C is the concentration (ppb) at time t, C_0 is the original concentration, e is the exponential constant, and k is a rate constant with units of time^{-1} that depends upon the mixing and flow rate through the lake.

We need to find values of the two constants, C_0 and k. In order to do this we use the two measurements that have been taken to generate two equations in the two unknowns as follows:

$$
\begin{aligned}
4.00 &= C_0 e^{-10k} \tag{4.7} \\
2.80 &= C_0 e^{-13k} \tag{4.8}
\end{aligned}
$$

To simplify these equations we take logs to base e.

$$\ln 4.00 = \ln C_0 - 10k \qquad (4.9)$$

$$\ln 2.80 = \ln C_0 - 13k \qquad (4.10)$$

Note that $\ln(e^z) = z$ from the definition of logarithm [eq. (3.1)]—think about it!

Subtracting equation (4.10) from equation (4.9) we have

$$\ln 4.00 - \ln 2.80 = -10k - (-13k)$$

and hence

$$k = \frac{\ln 4.00 - \ln 2.80}{3} = \frac{\ln(4/2.8)}{3} = 0.1189$$

Now we return to equation (4.7) in order to find the value of C_0 as follows:

$$4.00 = C_0 e^{-10k}$$

Multiply both sides by e^{10k} to give

$$4.00 e^{10k} = C_0 e^{-10k} e^{10k} = C_0 e^0 = C_0$$

Reverse the order and substitute k from equation (4.9). We then have

$$C_0 = 4.00 e^{10 \times 0.1189}$$
$$= 13.13 \text{ ppb}$$

At which point we take the offender to the cleaners! The original input was above the regulation threshold.

4.10 The Best Straight Line

4.10.1 Notation for Sums of Sequences

Mathematicians often work with sums of values. As you would expect, they have developed a notation to avoid writing down all the individual values. For example, if it was necessary to refer to the sum of the first n integer numbers, this would be expressed as

$$\sum_{i=1}^{n} i$$

which is mathematical shorthand for

$$1 + 2 + 3 + \cdots + n$$

Notice that we define the starting and finishing values below and above the Σ character and define the general term to be summed in an expression following it. This expression can be as complicated as necessary. The following defines the sum of the squares of all the even numbers from 2 to 100:

$$\sum_{i=1}^{50}(2i)^2$$

If we have a set of data such as $(x_1, y_1), (x_2, y_2), \ldots, (x_n, y_n)$, we could express the sum of the individual x values and the sum of the xy products as

$$\sum_1^n x_i \text{ and } \sum_1^n x_i y_i$$

respectively. If, however, we intend the sum to include all the values, as in the case above, it is not necessary to specify the starting and finishing values. The following equation defines the mean of all the x values:

$$\text{mean} = \frac{\sum x}{n}$$

4.10.2　Fitting the Best Straight Line

The values of x and y in the table below were measured during an experiment and are expected to be linearly related by the formula $y = mx + c$, where the parameters m and c are to be determined.

x	0	1	2	3	4	5
y	0.9	3.2	4.8	7.0	8.7	11.1

The best fit is calculated by minimizing the following sum of squares

$$SS = \sum_{i=1}^{n}[y_i - (mx_i + c)]^2$$

where y_i are the n observed values ($n = 6$ in this example) and $mx_i + c$ are the values predicted from the parameters m and c. It can be seen that when the fit is good, SS will be small; a large value for SS indicates a poor fit. In the case of a straight-line relationship it is possible, using calculus, to generate formulae for the best values of the two parameters (i.e., values that minimize SS) as follows:

$$m = \frac{\Sigma xy - \Sigma x \Sigma y / n}{\Sigma x^2 - \Sigma x \Sigma x / n}$$

$$c = \Sigma y - m \Sigma x / n$$

where n is the number of data points. This process is known as linear regression using the method of least squares.

We can perform the calculation by setting up a table as shown below. (On most pocket calculators and computers there is no need to do this because the calculation can be done automatically.)

	x	y	x^2	xy
	0	0.9	0	0
	1	3.2	1	3.2
	2	4.8	4	9.6
	3	7.0	9	21.0
	4	8.7	16	34.8
	5	11.1	25	55.5
Totals	15	35.7	55	124.1

Slope:

$$m = \frac{124.1 - (15)(35.7)/6}{55 - (15)(15)/6} = 1.986$$

Intercept:

$$c = 35.7/6 - 1.986(15)/6 = 0.995$$

The equation relating x and y is therefore

$$y = 1.986x + 0.995$$

4.11 The Michaelis Menton Equation

The Michaelis-Menton equation predicts the reaction velocity V in terms of the substrate concentration, $[S]$, as follows

$$V = \frac{V_{max}[S]}{[S] + K_m} \tag{4.11}$$

where V_{max} and K_m are parameters specific to the reaction and are usually determined from experiment and subsequent function fitting.

Given a table of data values $([S]_i, V_i)$ the problem is to define the values of the parameters V_{max} and K_m.

In the "olden days" this was difficult because there is no analytical solution that will generate the best values. Fitting curves was achieved either "by eye" or by transforming the data in some way so that the transformed data could be fitted. This usually implied that the transformation should result in a linear relationship so that the least squares fit (see section 4.10) could be used.

Scientists went to great lengths to linearize their data. In the case of fitting the Michaelis-Menton equation, two alternative transformations have been used, the Lineweaver-Burke transformation or the Eadie-Hofsee transformation.

4.11.1   The Lineweaver-Burke Transformation

We begin by multiplying the Michaelis-Menton equation (4.11) by $([S]+K_m)$ to give

$$([S] + K_m)V = V_{max}[S] \tag{4.12}$$

We then divide by $[S]$

$$\left(1 + \frac{K_m}{[S]}\right) V = V_{max}$$

divide by V and invert the equation

$$\frac{V_{max}}{V} = 1 + \frac{K_m}{[S]}$$

and finally, divide by V_{max}, which gives the Lineweaver-Burke equation:

$$\frac{1}{V} = \frac{1}{V_{max}} + \frac{K_m}{V_{max}} \times \frac{1}{[S]} \tag{4.13}$$

If we let $y = 1/V$ and $x = 1/[S]$, this is the equation of a straight line whose intercept is $1/V_{max}$ and slope is K_m/V_{max}. Thus, we can find the values of K_m and V_{max} by fitting the best (least squares) line.

4.11.2   The Eadie-Hofsee Transformation

Here, we multiply out the LHS of equation (4.12)

$$V[S] + VK_m = V_{max}[S]$$

divide through by $[S]$

$$V + \frac{VK_m}{[S]} = V_{max}$$

and rearrange to give the Eadie-Hofsee equation

$$V = V_{max} - K_m \left(\frac{V}{[S]}\right) \tag{4.14}$$

Plotting V against $V/[S]$ should give a straight line of slope $-K_m$ and intercept V_{max}.

4.11.3   Fitting the Parameters the Modern Way

There is no analytical formula for the best (least squares) parameter values, but we can calculate the sum of squares of the differences between the observed data and the values predicted by the equation using estimates for the values of V_{max} and K_m.

$$SS = \Sigma \left(V_i - \frac{V_{max}[S]_i}{[S]_i + K_m} \right)^2$$

It is then possible to optimize the fit by changing the values of the parameters so that the value of SS is minimized.

Initial estimates of the parameters V_{max} and K_m can be found as follows:

1. An estimate of V_{max} is easily gained from the data set by taking the maximum observed reaction velocity.

2. Estimating K_m is a little more difficult, but consideration of the equation provides a useful pointer as follows:

$$V = \frac{V_{max}[S]}{[S] + K_m}$$

$$\therefore K_m = \frac{V_{max}}{V}[S] - [S]$$

$$= [S]\left(\frac{V_{max}}{V} - 1\right)$$

From this equation we can see that when $V = V_{max}/2$, $K_m = [S]$, so that we can estimate K_m from the table by finding the value of $[S]$ when $V \approx V_{max}/2$.

The calculation is best done using a computer spreadsheet. An initial plot of the experimental data ($[s]$ on the x-axis, V on the y axis) will show up outlier points that should be checked. It will also be beneficial to plot the observed and predicted values and to try different parameter values in order to see how the fitted equation reacts. Progress can be made by manually changing the parameters to reduce the sum of squares value. This is a time-consuming process, but it is possible on most computer systems to optimize the fit automatically.

There are several advantages of this modern calculation:

- No transformation is necessary.

- Plotting the curve gives added insight, especially when the parameter values are changed manually. Doing so allows you to see the shape of the curve and its response to changes in parameter values.

- The calculation is easier.

- The calculations are carried out on the raw (untransformed) data so that the form of the errors is known.

- The errors on individual points can easily be weighted if necessary.

4.12 Graphs and Functions

It is almost always useful to depict data and functions in graphical form. Personal computers make life easy in this respect because data can be collected and displayed in spreadsheet form. You should *always* do this: it is much easier to detect anomalies from a graph than it is from a table, and it is always helpful to see the shape of your data.

4.12.1 Plotting Graphs

A scientist rarely needs anything other than an $x - y$ plot. Moreover, the initial look should be at an $x - y$ plot with no transformations and no lines joining the points. The reason is that transformations destroy the "shape" of the data, and drawing smooth lines to join experimental points can be misleading. In general there is no reason to believe that the fitted lines are an accurate representation of data that was *not* collected. Sometimes, however, if there are different data sets on the same graph, it is helpful to join the points within each set with straight lines so that the sets can be distinguished.

Spreadsheet graphics is a subject in itself and one which occupies more time and effort than it should! Remember that the capabilities provided in the popular software packages were designed to allow salespersons to impress their customers or supervisors. They were not designed particularly to help scientists. Most of the features are of little relevance to us.

After you have plotted and checked the data (and saved a backup copy), you can analyze the underlying trends and relationships. It may be that the relationships are known, in which case you should calculate the best fit to the data as in section 4.11.3. If the relationship is not known, consideration of the initial plot may reveal a shape that is familiar. Try fitting it: you may make discoveries, and even if you don't, you will learn a lot about the data. And it can be both rewarding and fun.

If the data set is very noisy, fitting it may not be possible, and a statistical investigation may prove to be the best way forward. Whatever the approach and subsequent analysis, it is our job as scientists to provide justification for any conclusions that we make. This will normally take the form of a statistical analysis, unless the relationship is obvious.

4.12.2 Shapes of Some Useful Functions

The following are a few sketches of relationships commonly used to model data. I have deliberately omitted scales on the graphs because the scale will depend upon the application.

The relationships (functions) are defined in terms of parameters a, $A(> 0)$, b, c, $k(> 0)$, and n. Parameters are constants involved in the function equations whose values are to be determined from experimental results. It may be possible to give a mechanistic interpretation of these parameters (e.g., V_{max} in the Michaelis-Menton equation is the maximum reaction velocity), and this is desirable, though not always possible.

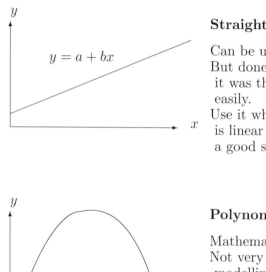

Straight line

Can be useful and often is.
But done to death because historically it was the only curve that could be fitted easily.
Use it when you know your data is linear or when you know nothing! It's a good start.

$y = a + bx$

Polynomial

Mathematical.
Not very useful in terms of mechanistic modelling because interpretation of the parameters is difficult or impossible.
Easy to deal with mathematically.

$y = a + bx + cx^2$

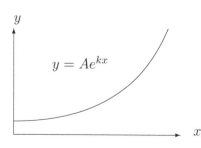

Exponential increase

The equation of growth.
Many applications in unrestricted growth, e.g., bacteria on agar plate, epidemics, nuclear reaction.
Parameters are meaningful.

$y = Ae^{kx}$

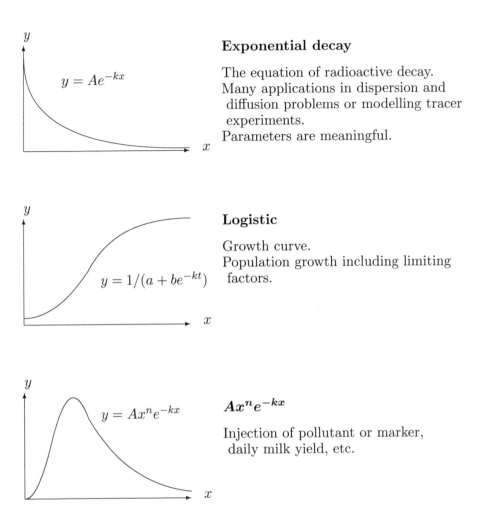

Exponential decay

The equation of radioactive decay.
Many applications in dispersion and
  diffusion problems or modelling tracer
  experiments.
Parameters are meaningful.

Logistic

Growth curve.
Population growth including limiting
  factors.

$Ax^n e^{-kx}$

Injection of pollutant or marker,
  daily milk yield, etc.

Normal frequency distribution

$$\frac{1}{\sqrt{2\pi\sigma^2}} \exp\left(\frac{(x-\mu)^2}{2\sigma^2}\right)$$

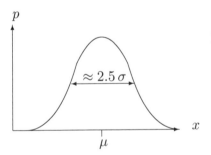

Chapter 5

Neat Tricks and Useful Solutions

I have chosen the topics in this chapter to introduce the more ambitious student to a few of the more taxing but useful areas of mathematical thinking. They are intended to stimulate the mind and to give more of an insight into the way mathematicians approach problems.

It is not intended that you, the student, should read and memorize these examples. It is hoped that you will occasionally look at one or another of them, whether for reference or fun or inspiration.

5.1 The Difference of Two Squares

For reasons that are not obvious, the expression $x^2 - a^2$ has great importance in school algebra. Students are asked to remember that

$$x^2 - a^2 = (x + a) \times (x - a)$$

When students ask (and they should ask), "Why does this relationship hold?" they are lucky if they get a reply. And if they do, it is most probably explained in the following way:

Because

$$
\begin{aligned}
(x + a)(x - a) &= x(x - a) + a(x - a) \\
&= x^2 - xa + ax - a^2 \\
&= x^2 - a^2
\end{aligned}
$$

Mathematically there is nothing wrong with the above, but

- it's the wrong way round, and

- it sheds no light on the nature of the problem.

Consider the following diagrams

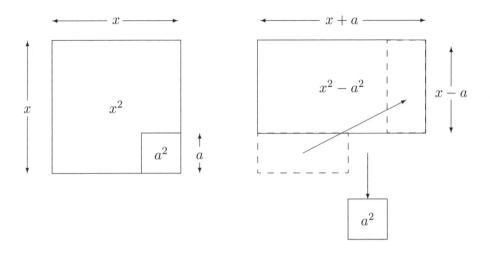

In the diagram on the left I have drawn two squares, one in the bottom right-hand corner of the other. The diagram on the right shows the effect of removing the smaller square, leaving an object with area $x^2 - a^2$, and rearranging what remains to make a rectangle with sides of length $x + a$ and $x - a$, and hence of area $(x + a) \times (x - a)$.

The moral of this story is that sometimes a picture is worth a thousand words. The equivalent mathematical viewpoint is that there are different ways of looking at a problem. In this case, a simple bit of geometry can clarify a difficult bit of algebra.

5.2 Mathematical Induction

The discovery of scientific rules often results from an observation that individual results follow some sort of pattern. For example, the following relationships

$$
\left.\begin{array}{rcrcl}
1 + 3 & = & 4 & = & 2^2 \\
1 + 3 + 5 & = & 9 & = & 3^2 \\
1 + 3 + 5 + 7 & = & 16 & = & 4^2
\end{array}\right\} \tag{5.1}
$$

may lead to the conclusion that

$$
1 + 3 + 5 + \cdots + (2n - 1) = n^2 \tag{5.2}
$$

where n is any +ve integer. Note that the observations in equation (5.1) only *suggest* that the general statement of equation (5.2) *may be true*. How can we prove it? One method that can be used is known as mathematical induction and takes the following form:

1. Show that the result is correct for a specific case ($n = 1$, say).

2. Assume that the result is true for some arbitrary value ($n = k$, say) and show that the result is also true for $n = k + 1$.

3. Combine the two previous results repeatedly to show that the formula is true for $n = 2, 3, \ldots$

Continuing the example mentioned above:

Step 1 We already know that the result is correct for $n = 2, 3, 4, \ldots$, from equation (5.1). In the case where $n = 1$ the result is trivial.

Step 2 Assume equation (5.2) is correct for $n = k$. Therefore,

$$\sum_{i=1}^{k} (2i - 1) = k^2 \tag{5.3}$$

for n = k + 1

$$
\begin{aligned}
\sum_{i=1}^{k+1} (2i - 1) &= \sum_{i=1}^{k} (2i - 1) && \text{the first } k \text{ terms} \\
&\quad + (2k + 1) && \text{the additional term} \\
&= k^2 + (2k + 1) && \text{from equation (5.3)} \\
&= (k + 1)^2
\end{aligned}
$$

which is the same as the result of applying equation (5.3) with k replaced by $k + 1$. Thus, if equation (5.2) holds for $n = k$, it must also hold for $n = k + 1$.

Step 3 Equation (5.2) must therefore hold for all +ve integers n, since

1. We know it to be true for $n = 1, 2, 3, 4$.

2. If it is true for $n = 4$, it must be true for $n = 5$.

3. Successive applications of step 2 for $n = 6, \ldots$, will include all the positive integers.

5.3   Pythagoras' Theorem

You will all be familiar with Pythagoras' theorem, $a^2 = b^2 + c^2$, which has many applications in geometry. The hypotenuse is the longest side of a right-angled triangle—the one opposite the right angle—and the theorem is often stated as

> The square of the hypotenuse is equal to the sum of the squares of the other two sides.

There are several proofs of the theorem, of which the following is probably the simplest.

Suppose we take a right-angled triangle and construct a square using four copies of it as in the diagram below.

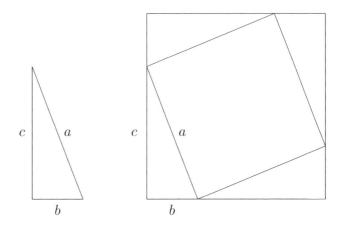

The area of the large square may be calculated using the lengths of its sides as

$$\text{Area} = (b + c)^2$$

Its area may also be calculated as the area of the smaller square plus four times the area of the triangle abc:

$$\text{Area} = a^2 + 4 \times \left(\frac{b \times c}{2} \right) = a^2 + 2bc$$

Now we can equate the two expressions for the area:

$$
\begin{aligned}
a^2 + 2bc &= (b + c)^2 \\
a^2 + 2bc &= b^2 + 2bc + c^2 \\
a^2 &= b^2 + c^2
\end{aligned}
$$

5.4   Pythagoras' Theorem Revisited

The following proof, discovered by a young Einstein (who was supposedly backward), is particularly stimulating and elegant.

Suppose we take a triangle:

Any old triangle will do for starters. Then draw a "similar" (that means the same shape) triangle with sides twice as long:

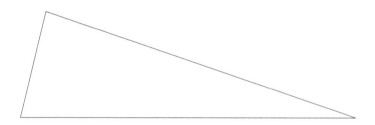

Is the area of the new triangle twice as large as the original, or if not, how much larger is it? In fact, it's easy to show that it's actually four times larger:

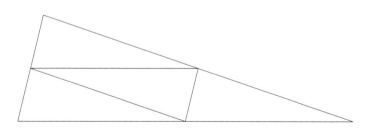

The original fits nicely into the new one four times exactly. (The middle one is upside down, but never mind that.)

If we draw a new (similar) triangle with sides three times as long, it's easy to show that the area is nine times the original.

So, for a similar triangle with sides twice as long as the original, the area is 4 times larger; for a triangle with sides three times as long the area is 9 times larger. If the sides are four times as long, the area is 16 times larger. You might like to draw it to convince yourself.

From the preceding arguments it looks as though the area of similar triangles is proportional to the square of the lengths of their sides. We can represent this by a bit of mathematics:

$$A = ka^2$$

where A is the area, a is the length of one of the sides (the equivalent one), and k is some constant. The value of this constant will depend upon the shape of the triangle and upon which side (a) we are comparing. To evaluate it you must measure the area (A) of a specific triangle and its corresponding side length (a); then $k = A/a^2$.

Actually, this rule applies to other shapes too: try proving it for rectangles. It's not so easy for other shapes but you can make most shapes out of a mixture of rectangles and triangles if you try!

So, for triangles of a given shape we have

$$A = ka^2$$

Now consider the triangle below, which contains a right angle at p. The line $p - s$ is a perpendicular.

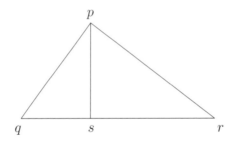

We have three similar triangles—pqs, rqp, and rps—whose corresponding sides are a, b, and c, as shown below.

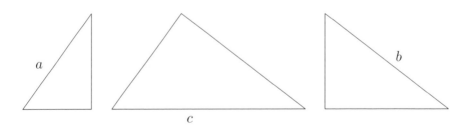

The areas will be ka^2, kc^2, and kb^2, respectively, where k is an unknown constant. In addition, the area of the large triangle is made up from the areas of the two smaller ones, so that

$$kc^2 = ka^2 + kb^2$$

Therefore, dividing throughout by k we have

$$c^2 = a^2 + b^2$$

Pythagoras knew this, but his proof was much more difficult. The young Einstein, struggling with the original version, felt that there ought to be an easier way.

5.5 Limits

The concept of a mathematical limit will strike you either as something intuitively obvious or as something so obscure as to be unimportant. Unfortunately for those of you in the latter group, limits are of fundamental importance in calculus, so that a basic understanding of them will be beneficial.

Consider the function

$$y = \frac{1}{x}$$

As x takes larger and larger positive values, y becomes smaller and smaller. It is also obvious that however large we make x, the value of y will never be negative. Thus, we can say that as x increases, y decreases towards zero, and therefore that the limiting value of y as x tends to infinity will be zero. We write this as

$$\lim_{x \to \infty} \frac{1}{x} = 0$$

In general we shall be concerned with limits where x approaches some specific value. As an example

$$\lim_{x \to 2} (3 - x) = 1$$

This example is so obvious as to be almost meaningless (what is the point of it all?). We can evaluate the limit simply by letting x take the limiting value and substituting in the function. Unfortunately, it is not always possible to do this. Consider:

$$\lim_{x \to 2} \frac{x^2 - 2x}{x - 2}$$

If we substitute the value $x = 2$ into this function the resulting expression is

$$\frac{0}{0}$$

which is indeterminate (not, in general, zero). However, if we factor the expression to give

$$\lim_{x \to 2} \frac{(x - 2)x}{(x - 2)} = x$$

we can see that, provided $x \neq 2$, we may cancel the factor $(x - 2)$ so that

$$\lim_{x \to 2} \frac{x^2 - 2x}{x - 2} = \lim_{x \to 2} x$$

and while x cannot take the value of 2 exactly, we may take x as close to the value 2 as we wish.

Thus, we write

$$\lim_{x \to 2} \frac{x^2 - 2x}{x - 2} = 2$$

It is most important to understand that in evaluating the limiting value, we have not simply calculated the function value at the limit: read this sentence again! In general, we can evaluate the function as close to the limit as we choose, but it may be impossible to calculate the value of the function at the limit. As close as we like—but not actually there. If this last paragraph isn't clear to you, read the section on limits again.

Example: The Limiting Value of $y = \frac{x^2 + x - 2}{x - 1}$ **as** $x \to 1$

$$
\begin{aligned}
\text{At } x &= 1, \; y = \frac{1 + 1 - 2}{1 - 1} = \frac{0}{0} \text{!!?} \\
\text{Now } y &= \frac{x^2 + x - 2}{x - 1} = \frac{(x - 1)(x + 2)}{x - 1} \\
\text{so that } \lim_{x \to 1} y &= \lim_{x \to 1} \frac{(x - 1)(x + 2)}{x - 1} \\
&= \lim_{x \to 1} (x + 2) \\
&= 3
\end{aligned}
$$

The concept of limiting values (limits) is the basis of calculus.

5.6   Trigonometry: Angles with a Difference

We use angles to describe the amount by which we rotate lines. For example, we can rotate a line by a full revolution, in which case it returns to its original position, or we can rotate it by a right angle, in which case it will be perpendicular to its original position. We normally measure angles in units of *degrees*, in which a full rotation corresponds to 360 degrees. Thus, a right angle is equivalent to 90 degrees, or $90°$. The following diagram contains an angle of approximately $45°$.

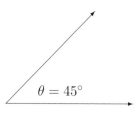

In order to measure angles we use a protractor, a semicircular template with the angles in degrees marked around the curved side.

5.6.1 Radians and Degrees

An alternative method of measuring angles is to measure the length of the arc subtended by the angle and divide this by the radius. In this method the angle is measured in units of *radians*. A full rotation ($360°$) corresponds to 2π radians.

In science it is often more convenient to use radians than degrees so it is helpful to understand the difference. Your calculator will work with either unit, but if it is working in radians and you are measuring in degrees, you will get the wrong answers, so watch out!

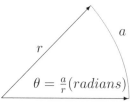

The following table shows equivalent angles in degrees and radians

Degrees (°)	0	45	57.3	90	114.6	180	360
Radians	0	$\pi/4$	1	$\pi/2$	2	π	2π

5.6.2 Trigonometric Ratios: sine, cosine, tangent

The trigonometric functions can all be associated with ratios of the lengths of the various sides of a right-angled triangle. Consider the following right angled-triangle containing the angle θ.

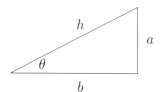

The values of the various trigonometrical functions (sine, cosine and tangent) are defined as follows:

$$\sin \theta = a/h = \text{opposite/hypotenuse}$$
$$\cos \theta = b/h = \text{adjacent/opposite}$$
$$\tan \theta = a/b = \text{opposite/adjacent}$$

and are normally written in the above short forms.

When the angles are measured in radians, notice that as $\theta \to 0$, $\sin \theta \to \theta$ and $\cos \theta \to 1$, because $a \to h\theta$ and $b \to h$. See section 5.5 concerning limiting values.

Using the definitions above, together with Pythagoras' theorem, other useful relationships can be found, such as

$$\sin^2 \theta + \cos^2 \theta = 1$$

and

$$\tan \theta = \frac{\sin \theta}{\cos \theta}$$

In order to convince yourself that you understand the trigonometric ratios, prove the two relationships.

Application: Radiation on a Surface

The intensity of radiation is usually measured in terms of the amount of radiation falling on an area perpendicular to the direction of radiation. Thus, radiation of 100 W m^{-2} implies that an area of 1 m^2 at right angles to the radiation would receive 100 W of radiative energy.

However, on a surface which is not at right angles to the direction—for example, the solar panel on your house roof, should you decide to go green:

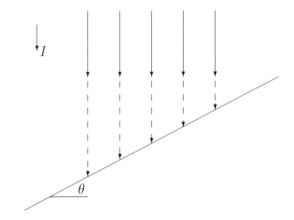

In this case the intensity on the inclined surface is

$$\text{radiation intensity} \times \frac{\text{area at right angle}}{\text{area of shadow}} = I \times \cos\theta$$

Application: What Force on the Biceps?

The resolution of forces within the bone and muscle system requires the application of a little trigonometry. The following diagram is a simple representation of the arm. Here, we are interested in calculating the force within the biceps that will be necessary to lift a weight of mass m. The lengths involved are denoted by b, h, r, and d, while F and mg are the forces.

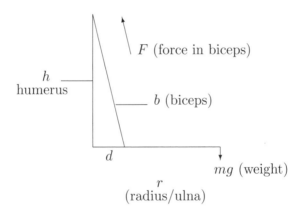

The force exerted by the biceps will act along its length, but we will need to find the vertical component in order to balance the moments about the elbow of the weight $(mg \times r)$ and the upward force exerted by the muscle a distance (d) from the elbow.

If θ is the angle between the humerus and the biceps, the vertical component of F will be $F\cos\theta$, so that

$$F\cos\theta \times d = mg \times r$$

Rearranging gives

$$F = \frac{mgr}{d\cos\theta}$$

and we can find $\cos\theta$ as follows:

$$\cos\theta = \frac{h}{\sqrt{h^2 + d^2}}$$

The necessary force is given by

$$F = \frac{mgr\sqrt{h^2 + d^2}}{dh} = \frac{mgrb}{dh}$$

5.7   Numerical Calculations

Before the advent of the computer dramatically reduced the time and effort involved in long calculations, scientists were restricted to the solution of relatively simple numerical problems. This meant that efforts were concentrated on developing analytical methods, the kind of processes used thus far in this book. Large arithmetic calculations were impossible with only pencil and paper, and so problems were simplified by making gross assumptions: you may already have come across frictionless planes, spherical cows, and similar improbable phenomena. The evolution of the computer, together with the development of numerical analysis, have provided the opportunity to tackle some of the more realistic problems that don't fit into the idealized world of classical mathematics.

The application of numerical methods to problems in life sciences has been tenuous at best, but for those of you who are prepared to use this approach the rewards could be enormous.

5.7.1   Iteration

Iteration is a method by which we can calculate numerical solutions to problems that cannot be solved by analytical means. The method consists of making an educated guess and then devising a process for successively improving the solution until we are satisfied with the accuracy obtained.

A method that Newton used to calculate square roots is a good, though specific example. It is convenient to describe this method in terms of an *algorithm*, which is a fancy word for a recipe. In order to calculate x, which is the square root of y, the algorithm consists of a sequence of steps as follows:

1. Make a guess x_1 at the square root.

2. Calculate an improved value x_2 as follows

$$x_2 = \frac{x_1 + y/x_1}{2}$$

   so that x_2 is the average of x_1 and (y/x_1). It should be apparent that if the value of x_1 is too large, then the value of (y/x_1) will be too small, and hence the average value will be a better estimate.

3. If x_2 and x_1 are sufficiently close together, then stop and set $x = x_2$; otherwise, set $x_1 = x_2$ and go to step 2.

The process consists of three parts: an initial estimate, a repetitive process which produces successively better estimates, and a test which terminates the process when the required accuracy has been reached. Finding the initial estimate may be a problem, but the value 1.0 will generate a result.

A starting value of zero is not good. (Why?) Try the process out using a spreadsheet with values of y for which you know the square root.

5.7.2 The Method of Iteration

Having derived a relationship of the form

$$y = f(x)$$

we will occasionally need to answer questions like "What is the value of x when y is 3.5?" The answer to this question may involve a simple reorganization of the formula, as in section 1.3.1, but often it will be impossible to re-arrange the expression to give x in terms of y. Plotting a graph may provide a solution, but it is a time-consuming exercise, especially if accuracy is required. It is often a relatively simple matter to solve such problems numerically, and there are many standard methods available on modern computer systems.

Suppose, for example, that we have performed an experiment in which we plotted the concentration (c) of a chemical against time (t) and found that the concentration could be expressed in terms of the following function of time:

$$c = t^2 e^{-t}$$

Recourse to the pocket calculator generates the following points:

t	0	1	2	3	4	5	6	7
c	0	.368	.541	.448	.293	.168	.089	.045

from which we are able to sketch the curve plotted below.

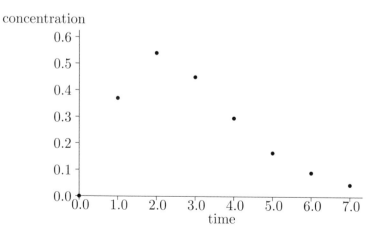

A question we might want to ask is, "When does the concentration reach 25 percent?" This is, in effect, the same as asking for the value of t that satisfies the equation

$$0.25 = t^2 e^{-t} \tag{5.4}$$

The solution of this equation or, more generally, of the transformed equation

$$t^2 e^{-t} - 0.25 = 0 \tag{5.5}$$

is the subject of this section. It will be apparent that a direct solution of equation (5.5) is not obvious.

A technique which is often worth exploiting is to rearrange the equation so that t is expressed as a function of itself. One such expression can be derived as follows:

$$
\begin{aligned}
0.25 e^t &= t^2 \\
e^t &= 4t^2 \\
t &= \ln(4t^2) \tag{5.6}
\end{aligned}
$$

The object of this expression is to give a method of generating successive approximations to the value of t that we hope will converge to the required value. This method is known as the method of iteration. We could express these successive approximations as

$$t_{n+1} = \ln(4t_n^2) \tag{5.7}$$

indicating that we calculate a new value t_{n+1} by substituting the current value t_n into equation (5.6). From our graph we could take an initial approximation for t_0 to be 1.0 (a guess of $t_0 = 0$ would be foolish!). Then using equation (5.7) we can generate successive approximations as follows:

n	t_n
0	1
1	1.386
2	2.040
3	2.812
4	3.454
5	3.865
6	4.090
7	4.204
8	4.258
9	4.284

n	t_n
10	4.296
11	4.302
12	4.304
13	4.306
14	4.306

It can be seen that the successive estimates converge and the required value is 4.306 ± 0.001. We can stop the process when the difference between successive iterations has reached the required accuracy (0.001 in this case). A look at our plot will reveal that there is indeed a concentration of 0.25 in the region of $t = 4.3$, but we would probably have preferred to find the value in the region of $t = 0.7$! There are often several answers to a problem, and we must be careful that we have calculated the correct one.

How then do we get an accurate value for this other solution? A different rearrangement of equation (5.4) may give the required result, so let us try

$$t_{n+1} = \sqrt{0.25 \exp(t_n)} \tag{5.8}$$

with an initial guess of 1.0. This generates the following sequence:

n	t_n
0	1.000
1	0.824
2	0.755
3	0.729
4	0.720
5	0.717
6	0.715
7	0.715

This time we are lucky that the result converges to the required value.

It is possible to predict whether a particular iterative expression will converge, but in practice this can be difficult. It is often easier to calculate the first few values in the sequence; then, if the values don't converge we can try a different arrangement of the expression. Many problems lend themselves to this kind of solution.

5.7.3 The Method of Bisection

Sometimes it may be difficult to rearrange an equation into a convergent iterative process. One reliable method of solving an equation of the type

$$g(x) = 0$$

is to find values of x which bracket the root and successively halve the interval until the two bracketing values are sufficiently close. This method may be expressed as an algorithm as follows:-

1. Find x_1, x_2 so that $g(x_1) \times g(x_2) < 0$.

2. If $|x_2 - x_1| <$ tolerance then stop.

3. Calculate $x_3 = 0.5 \times (x_1 + x_2)$.

4. If $g(x_3) \times g(x_1) < 0$ set $x_2 = x_3$; otherwise set $x_1 = x_3$.

5. Go to step 2.

This method is known as the method of bisection.

There are many other methods for finding the zeros of functions. If it is necessary to find an efficient solution, you should consult the relevant numerical methods textbooks and the tools available on your computer system. The *Solver* "add-in" within Excel is one such tool.

Chapter 6

Differential Calculus

6.1 Introduction

The word *calculus* is Latin for "pebble." Early forms of the abacus used pebbles in grooves marked out in sand; hence, the association. In its generic form the word means any method of calculation. Therefore, you have been making use of calculus for years!

However, in more recent times *the calculus* has come to mean the branch of mathematics that is concerned with the behavior of dynamic systems, that is, with systems in which objects move or change, like all living things— bacteria, cows, and humans! It was developed by Fermat, Newton, and others in order to study the motion of planets, pendulums ... and falling apples? Current applications include the modelling of plant and animal development and aspects of population growth, epidemiology, and so forth. Most computer models are based on the methods of calculus, though they use numerical approximations in order to solve the (much) more complex equations necessary to describe these systems. The ability to construct differential equations that define such systems will allow you to make use of the many computer packages that can produce solutions.

Calculus comprises two processes: *differentiation*, in which we know the equations defining the state of a system and use them to work out the rate at which the system will change as its independent variables change, and *integration*, in which we know the equations defining the rate of change and use them to predict the state of the system at specific values of its variables. Many people, especially west of the Atlantic, refer to integration as *anti-differentiation*—a hideous but usefully accurate term!

I will begin by describing differential calculus because differentiation can be defined using one formula (though working with it can be extremely tedious), while integration is more of an art form, relying to a large extent on guesswork and experience gained from differentiation.

6.2   What Is Differentiation?

The following math-speak introduces some complicated notation. Don't panic! Just accept it for what it is—you will soon understand.

If we know a relationship $y = f(x)$, it is often possible to derive a formula that defines the slope of its graph at any point x. This formula, which we denote by dy/dx or $f'(x)$, is variously called the *derivative* or the *differential* or the *differential coefficient* of the function $y = f(x)$. The process that we go through in order to find it is known (to mathematicians) as *differentiation*.

Differentiation is a word which causes me, and (I suspect) many of you, great difficulty because its colloquial meaning gives little clue to its mathematical use. I am more at home with the word *differential*, because I know that the differential on a car allows the right- and left-hand wheels to turn at different speeds when cornering. I am also familiar with the term *differential* referring to bicycles, where the differential is the ratio of wheel velocity to pedal or crank velocity.

The differential on a bicycle is the ratio of the number of teeth on the chainwheel (the one connected to the pedals) to the number of teeth on the rear wheel sprocket; it tells you the relative speed (rpm) of the wheels corresponding to the speed (rpm) of the pedals. Thus, if the chainwheel has 48 teeth and the rear wheel has 16 teeth, the differential is 3, because for each turn of the pedals the wheels go round three times.

Also, if we plot the angular speed of the wheels against the speed of the pedals, we will see a straight-line graph whose slope is 3. The differential coefficient is the slope of such a graph.

Thus, the derivative of the function defining wheel velocity in terms of pedal velocity is the same as the differential, which is also the slope of the graph of wheel angular velocity against pedal velocity.

Differential coefficients are quantities or expressions that determine the relative change in a variable as its independent variable changes. Usually the independent variable will be time. Thus, the growth rate is the change in mass divided by the corresponding change in time. We would refer to this as dm/dt. Acceleration or the rate of change of velocity may also be expressed as a differential coefficient (dv/dt)—in other words the relative change in velocity as time changes. The independent variable will not necessarily always be time. For example, we could ask what the relative change in the area of a circle is as the radius changes; in this case we would require an expression for dA/dr.

In general, if we can define a function which specifies the size, position, concentration, etc., of an object at a given time. Then differentiation will enable us to derive equivalent functions for the growth rate, velocity, reaction rate, etc., of the object.

6.3 Distance and Velocity

Legend would have us believe that, one day while lying on his back in the orchard, Newton invented the formula

$$s = \frac{1}{2}gt^2$$

which predicts the distance fallen s by a particle at time t from rest. (The term g is the gravitational constant, which in SI units is 9.81 m s^{-2}.) Note that this equation ignores air resistance, but this inspiration probably came to Newton on one of those sultry, airless days!

We can use this equation to calculate s every quarter of a second during the first 2 seconds of its fall, to give the following table:

t	0.00	0.25	0.50	0.75	1.00	1.25	1.50	1.75	2.00
s	0.00	0.31	1.23	2.76	4.91	7.66	11.04	15.02	19.62

from which we can plot a graph of s against t.

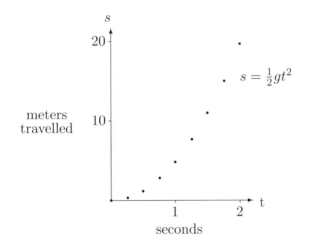

Figure 6.1: Distance against Time

We could fill in the gaps, either by calculating further values or by fitting a smooth curve through the existing points. If we do fit a curve through the points, we are approximating, but the alternative involves a *lot* of calculation!

Figure 6.1 allows us to see how far the particle has dropped at any given time, or it would if we filled in the gaps. However, this in itself doesn't provide any more information than the original formula.

Differential calculus is a mathematical technique that allows us to find an expression (equation) defining the speed of the particle at any specified time.

6.3.1   Average Velocity

The average velocity of the particle over a given time interval is given by the distance travelled divided by the time taken. This is an easy calculation. For example, if we want to know the average velocity over the first second, $\bar{v}_{0,1}$, we can calculate it as follows:

$$\bar{v}_{0,1} = \frac{4.91 - 0.00}{1 - 0} = 4.91 \text{ m s}^{-1} \tag{6.1}$$

or if we wanted to calculate the average velocity between 1 and 2 seconds:

$$\bar{v}_{1,2} = \frac{19.62 - 4.91}{2 - 1} = 14.71 \text{ m s}^{-1} \tag{6.2}$$

You will observe that each answer is the slope of the chord joining the two points on the graph corresponding to the specified time interval.

6.3.2   Instantaneous Velocity

The instantaneous velocity is given by the slope of the tangent to the graph at a particular time, so we could calculate the velocity by drawing the tangent to the graph and then measuring its slope. This might be fun (?) but is certainly tedious if we are expected to produce accurate results or have to do it more than once. There must be an easier way!

Suppose we are required to find the velocity of the particle 1 second after it has been dropped.

One estimate of this velocity would be the average calculated over the interval 1 to 2 seconds. Observation says that a better estimate can be made using the average over the interval 1 to 1.5 seconds, and a better one still using the average over 1 to 1.25 seconds. Table 6.1 shows the results of using smaller and smaller intervals based on the time interval beginning at 1 second.

Table 6.1: Average Velocity Just after 1 Second

time interval (t)	$\bar{v}$ (m s^{-1})
1.0000 to 2.0000	14.7150
1.0000 to 1.5000	12.2625
1.0000 to 1.2500	11.0363
1.0000 to 1.1000	10.3005
1.0000 to 1.0100	9.8590
1.0000 to 1.0010	9.8149
1.0000 to 1.0001	9.8105

An alternative approach is to calculate the average velocity over small time intervals just before 1 second, as in table 6.2. It can be seen from these tables that the average velocity over the smaller intervals is converging to the value 9.81.

Table 6.2: Average Velocity Just before 1 Second

time interval (t)	$\bar{v}$ (m s^{-1})
0.0000 to 1.0000	4.9050
0.5000 to 1.0000	7.3575
0.7500 to 1.0000	8.5838
0.9000 to 1.0000	9.3195
0.9900 to 1.0000	9.7610
0.9990 to 1.0000	9.8051
0.9999 to 1.0000	9.8096

More elaborate procedures that converge more quickly can be devised, using time limits straddling the specific time; however, the important thing to notice is that the values do converge to a limit, and that this limit is independent of the approach.

The numerical method used above is now a standard method for evaluating *rates*, but in Newton's time, and for a considerable period afterwards (until the 1930s) the arithmetic was too time consuming.

Newton and his contemporaries didn't have the advantage of modern computers to solve such problems, so they looked for more convenient analytical methods. This was where "the calculus" was conceived. The calculus solution follows a process similar to the one above but avoids specific numbers in order to find a general formula for the velocity at any value of t.

Earlier, we estimated the instantaneous velocity at time t by calculating the average velocity over a small time interval beginning at t. The mathematician's way of saying this is to look at the time interval between t and $t+\delta t$, where t is any value of time and δt is a small interval. (Mathematicians use the Greek letter δ, delta, as shorthand for "a small amount of . . .")

In order to calculate the average velocity we need to calculate the distance travelled (δs) during the time interval δt.

$$\delta s = \frac{1}{2}g(t + \delta t)^2 - \frac{1}{2}gt^2 \qquad (6.3)$$

The estimate of velocity at time t is given by

$$
\begin{aligned}
v(t) = \frac{\delta s}{\delta t} &= \frac{\frac{1}{2}g(t + \delta t)^2 - \frac{1}{2}gt^2}{\delta t} \\
&= \frac{g\{(t + \delta t)^2 - t^2\}}{2\delta t} \\
&= \frac{g\{t^2 + 2t\delta t + \delta t^2 - t^2\}}{2\delta t} \\
&= \frac{g\{2t\delta t + \delta t^2\}}{2\delta t} \\
&= gt + \frac{1}{2}g\delta t
\end{aligned}
$$

This is a general expression for equations 6.1 and 6.2, giving the average velocity between t and $t + \delta t$ seconds.

Now we know that the best estimate of the instantaneous velocity is calculated when the time interval is made as small as possible. The mathematician's way of saying this is when $\delta t \to 0$ (i.e., the time interval approaches zero).

If we let $\delta t \to 0$, the second term in the equation disappears (because we can make it small enough to ignore), and we are left with

$$
\text{velocity} = \lim_{\delta t \to 0} \frac{\delta s}{\delta t} = gt
$$

As usual, mathematicians have a shorthand way of writing

$$
\lim_{\delta t \to 0} \frac{\delta s}{\delta t};
$$

they write this as

$$
\frac{ds}{dt}
$$

and call it the *differential coefficient* of s with respect to t. So we have

$$
\text{velocity} = \frac{ds}{dt} = gt \tag{6.4}
$$

It can be seen that when $t = 1$, $ds/dt = 9.81$, which agrees with our previous calculation. The value of the velocity at $t = 0$ is also easy to calculate ($= 0$) and agrees with intuition and our graph (fig. 6.1). Thus, we have some confidence in equation (6.4) and are now in a position to be able to calculate the velocity at any instant.

Pause for meditation: If the differential of $gt^2/2$ is gt, must the anti-differential of gt be something to do with $gt^2/2$?

6.4 The Differential Coefficient of Any Function

The calculation of velocity carried out in the previous section can be generalized to provide an expression for the differential coefficient of any function.

Suppose we are given the relationship $y = f(x)$, where $f(x)$ is some specified function of x.

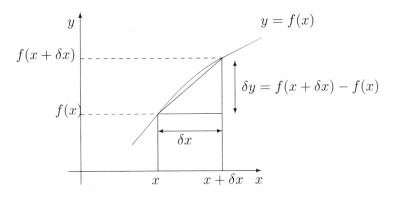

Figure 6.2: Estimating the Slope of $y = f(x)$ at $(x, f(x))$

In this case we look at two values of the function at x and $x + \delta x$. We can calculate the corresponding y values as $f(x)$ and $f(x+\delta x)$ and use them to calculate the change δy in y corresponding to the change δx in x.

The slope of the curve $y = f(x)$ at the point (x, y) is now given by

$$\frac{dy}{dx} = \lim_{\delta x \to 0} \frac{\delta y}{\delta x} = \lim_{\delta x \to 0} \frac{f(x + \delta x) - f(x)}{\delta x} \tag{6.5}$$

Equation (6.5) defines the *differential coefficient* (*derivative*) of y with respect to x, and is the basis of differential calculus: all of what follows is dependent upon it. Working with this equation can be extremely tedious and time consuming, as you will see from some of the following examples. However, tables of standard derivatives exist, and there are methods and shortcuts that make life easier. We will develop some of these in the following sections. If you can, though, you should try to understand what differentiation is about at least once: there is no substitute for knowing what is going on.

Example: The Differential Coefficient of the Function $y = 3x$

$$\begin{aligned}
\frac{dy}{dx} &= \lim_{\delta x \to 0} \frac{3 \times (x + \delta x) - 3 \times (x)}{\delta x} \\
&= \lim_{\delta x \to 0} \frac{3x + 3\delta x - 3x}{\delta x} \\
&= \lim_{\delta x \to 0} \frac{3\delta x}{\delta x} \\
&= 3
\end{aligned}$$

This answer tells us that the slope of the function $y = 3x$ is 3, irrespective of the value of x—i.e., it is a straight line. This should come as no surprise to you, but if you understand it, you can give yourself a pat on the back. The mathematics looks horrid, but the ideas aren't too bad if you just dig your way through the jargon. In general you can ignore the jargon, but sometimes it is helpful to be precise.

Example: The Derivative of the Function $y = x^2 + 3x + 2$

$$\begin{aligned}
\frac{dy}{dx} &= \lim_{\delta x \to 0} \frac{[(x + \delta x)^2 + 3(x + \delta x) + 2] - [x^2 + 3x + 2]}{\delta x} \\
&= \lim_{\delta x \to 0} \frac{x^2 + 2x\delta x + \delta x^2 + 3x + 3\delta x + 2 - x^2 - 3x - 2}{\delta x} \\
&= \lim_{\delta x \to 0} \frac{(2x + \delta x + 3)\delta x}{\delta x} \\
&= \lim_{\delta x \to 0} 2x + 3 + \delta x \\
&= 2x + 3
\end{aligned}$$

Example: The Derivative of $\sqrt{x}$

$$\frac{d}{dx}\sqrt{x} = \frac{d}{dx} x^{\frac{1}{2}}$$

In this case it is convenient to express the estimate of slope in a different way. An alternative form of the equation is

$$\frac{\delta y}{\delta x} = \frac{f(q) - f(x)}{q - x} \qquad \text{think of } q \text{ as } x + \delta x$$

so that we have

$$\frac{\delta y}{\delta x} = \frac{q^{\frac{1}{2}} - x^{\frac{1}{2}}}{q - x}$$

$$= \frac{q^{\frac{1}{2}} - x^{\frac{1}{2}}}{(q^{\frac{1}{2}} - x^{\frac{1}{2}})(q^{\frac{1}{2}} + x^{\frac{1}{2}})} \qquad [\text{since } a^2 - b^2 = (a-b)(a+b)]$$

$$\therefore \frac{d}{dx} x^{\frac{1}{2}} = \lim_{q \to x} \frac{1}{q^{\frac{1}{2}} + x^{\frac{1}{2}}}$$

$$= \frac{1}{2x^{\frac{1}{2}}} = \frac{1}{2} x^{-\frac{1}{2}}$$

$$\therefore \frac{d}{dx} x^{\frac{1}{2}} = \frac{1}{2} x^{-\frac{1}{2}}$$

6.5 Differentiability

A function $y(x)$ is said to be differentiable if the change δy in y, corresponding to a change δx in x, can be made arbitrarily small by choosing δx as small as we wish.

Most functions which describe physical situations are differentiable. Such functions are often referred to as being "well behaved," "smooth," or "continuous." The following is an example of a function which is not differentiable:

$$x = \begin{cases} 1 & \text{if } x > 0 \\ -1 & \text{otherwise} \end{cases}$$

This function is not differentiable at $x = 0$ and is thus called *discontinuous*.

The physical significance of the above is that a function will be differentiable provided it has no sharp corners or values which are infinitely large.

6.6 Evaluation of Some Standard Derivatives

As noted earlier, tables of derivatives are available, and modern computer software with symbolic algebra capabilities is able to solve many differentiation problems. Nevertheless, much can be done with a small selection of derivatives, and in this section I will show the evaluation of the most important of these. The derivation of these differential coefficients is of no great importance in itself, though some of you may like to know how the mathematical arguments go. However, it will be useful to follow the arguments through at least once, if only to gain insight into the methods of mathematical proof.

The Derivative of a Constant

$$\text{Let } y = \text{constant}$$

$$\frac{dy}{dx} = \lim_{\delta x \to 0} \frac{f(x + \delta x) - f(x)}{\delta x}$$

$$= \lim_{\delta x \to 0} \frac{A - A}{\delta x} = 0$$

$$\frac{d}{dx} \text{ constant} = 0$$

This result is intuitive, since the slope (and hence the derivative) of the line $y = $ constant, is zero.

The Derivative of x^n

$$\frac{d}{dx} x^n = \lim_{\delta x \to 0} \frac{(x + \delta x)^n - x^n}{\delta x}$$

$$(x + \delta x)^n = x^n + nx^{n-1}\delta x + \frac{n(n-1)}{2!}x^{n-2}\delta x^2 + \cdots + nx\delta x^{n-1} + \delta x^n$$

$$= x^n + nx^{n-1}\delta x + 0(\delta x^2)$$

The expression $O(\delta x^2)$ is a shorthand way of saying "all the remaining terms which involve powers of the Order δx^2 or greater."

$$\frac{d}{dx} x^n = \lim_{\delta x \to 0} \frac{x^n + nx^{n-1}\delta x + O(\delta x^2) - x^n}{\delta x}$$

$$= \lim_{\delta x \to 0} nx^{n-1} + O(\delta x)$$

$$= nx^{n-1} \qquad \text{since all terms } O(\delta x) \to 0$$

$$\text{Thus } \frac{d}{dx} x^n = nx^{n-1}$$

The Derivatives of $\sin x$ and $\cos x$

$$\frac{d}{dx} \sin x = \lim_{\delta x \to 0} \frac{\sin(x + \delta x) - \sin x}{\delta x}$$

$$= \lim_{\delta x \to 0} \frac{\sin x \cos \delta x + \cos x \sin \delta x - \sin x}{\delta x} \qquad (6.6)$$

because, by reference to equation (8.5) in the chapter on matrix algebra

$$\sin(\theta + \phi) = \sin \theta \cos \phi + \cos \theta \sin \phi$$

Consider the following diagram:

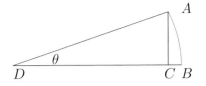

If θ is measured in radians,

$$\theta = \frac{AB}{BD} = \frac{\text{length of arc}}{\text{radius}}$$

$$\sin \theta = \frac{AC}{BD}$$

From inspection: as $\theta \to 0$ $AC \to AB$

$$\therefore \lim_{\theta \to 0} \sin \theta = \theta$$

$$\cos \theta = \frac{DC}{AD} = \frac{DC}{DB}$$

From inspection: as $\theta \to 0$ $DC \to DB$

$$\therefore \lim_{\theta \to 0} \cos \theta = 1$$

Using these two limits we can rewrite equation (6.6):

$$\frac{d}{dx} \sin x = \lim_{\delta x \to 0} \frac{\sin x . 1 + \cos x \delta x - \sin x}{\delta x}$$

$$= \lim_{\delta x \to 0} \cos x$$

$$\frac{d}{dx} \sin x = \cos x$$

By similar arguments we can find the derivative of $\cos x$

$$\frac{d}{dx} \cos x = -\sin x$$

The Derivative of a Constant Times a Function of x

$$\frac{d}{dx} af(x) = \lim_{\delta x \to 0} \frac{af(x + \delta x) - af(x)}{\delta x}$$

$$= \lim_{\delta x \to 0} \frac{a[f(x + \delta x) - f(x)]}{\delta x}$$

$$= a \lim_{\delta x \to 0} \frac{f(x + \delta x) - f(x)}{\delta x}$$

$$\therefore \frac{d}{dx} af(x) = a \frac{d}{dx} f(x)$$

Example: The Derivative of $3\sin x$

$$= 3\frac{d}{dx}\sin x = 3\cos x$$

Example: The Derivative of $4x^2$

$$\frac{d}{dx}4x^2 = 4\frac{d}{dx}x^2$$
$$= 4 \times 2x = 8x$$

Food for thought: If the derivative of $4x^2$ is $8x$, then so is the derivative of $4x^2 + c$ (where c is an unknown constant). So could the integral (anti-differential) of $8x$ be $4x^2 + c$?

The Derivative of e^x
Consider the function a^x, where a is an arbitrary constant.

$$\frac{d}{dx}a^x = \lim_{\delta x \to 0}\frac{a^{x+\delta x} - a^x}{\delta x}$$
$$= \lim_{\delta x \to 0}a^x\frac{(a^{\delta x} - 1)}{\delta x}$$
$$= a^x \times \lim_{\delta x \to 0}\frac{a^{\delta x} - 1}{\delta x}$$

The limit is difficult to evaluate, but we can show that it exists if we look at a few explicit examples.

$$\text{Consider } \frac{d}{dx}2^x = 2^x \times \lim_{\delta x \to 0}\frac{2^{\delta x} - 1}{\delta x}$$

δx	$\frac{2^{\delta x}-1}{\delta x}$
0.1	0.717735
0.01	0.695555
0.001	0.693387
0.0001	0.693169

It can be seen that as $\delta x \to 0$ the value of the expression converges to 0.693147.

$$\text{Similarly } \frac{d}{dx}2.5^x \approx 2.5^x \times 0.916291$$
$$\text{and } \frac{d}{dx}3^x \approx 3^x \times 1.098612$$

It appears that there should be some value of a such that $\frac{d}{dx}a^x = a^x$ and that this value must lie somewhere between 2.5 and 3.0. In fact, this number turns out to be the value $2.718282\ldots$, which is the exponential constant (e) and is sometimes referred to as Euler's number after the scientist who discovered it. Thus,

$$\frac{d}{dx}e^x = e^x \tag{6.7}$$

The exponential term e^x is often expressed as a function $\exp(x)$; the two are interchangeable and equivalent. The exponential function is an infinite series as follows:

$$\exp(x) = 1 + x + \frac{x^2}{2!} + \frac{x^3}{3!} + \cdots + \frac{x^n}{n!} + \cdots$$

$$= \sum_{i=0}^{\infty} \frac{x^i}{i!}$$

where $n!$ (referred to as "factorial n") is calculated as $n(n-1)(n-2)\cdots(2)(1)$, for example, $3! = (3)(2)(1) = 6$.

The derivative of $\exp(x)$ is therefore the sum of the derivatives of the individual terms in the series:

$$\exp(x) = 1 + x + \frac{x^2}{2!} + \frac{x^3}{3!} + \cdots$$

$$\frac{d}{dx}\exp(x) = 0 + 1 + \frac{2x}{2!} + \frac{3x^2}{3!} + \cdots$$

$$= 0 + 1 + x + \frac{x^2}{2!} + \cdots$$

$$\frac{d}{dx}\exp(x) = \exp(x) \qquad \text{since the RHS} = \sum_{i=0}^{\infty} \frac{x^i}{i!}$$

The Derivatives of $\ln x$ and a^x

$$\text{If } y = \ln x$$

$$x = e^y$$

$$\text{so that } \frac{dx}{dy} = e^y$$

$$\text{but } \frac{dy}{dx}\frac{dx}{dy} = 1$$

$$\therefore \frac{dy}{dx} = \frac{1}{\left(\frac{dx}{dy}\right)} = \frac{1}{e^y} = \frac{1}{x}$$

$$\frac{d}{dx}\ln x = \frac{1}{x}$$

We have seen that the derivative of a^x (where a is a constant) is difficult to evaluate from first principles. The problem is relatively simple now, however, since we can use logarithms. The derivative is found as follows:

$$\text{if } y = a^x$$
$$\ln y = x \ln a \qquad\qquad (\ln \equiv log_e)$$
$$\therefore x = \frac{\ln y}{\ln a}$$
$$\text{and } \frac{dx}{dy} = \frac{1}{\ln a}\left(\frac{d}{dx}\ln y\right) = \frac{1}{y \ln a}$$
$$\therefore \frac{dy}{dx} = y \ln a$$
$$\frac{d}{dx}(a^x) = a^x \ln a$$

6.7   Derivatives Involving Two Functions

The calculation of complicated derivatives can be extremely tedious and difficult, especially if we have to resort to first principles using equation (6.5). It will be useful to develop a few shortcuts and tools to ease the process. The following is a formal treatment showing how the rules for differentiating the sum, product, and quotient of two functions are developed. While the rules are important, their derivation may be found tedious and may be skipped over by the faint-hearted. The applications and examples should be mastered, however.

Consider two differentiable functions of x: $u = u(x)$ and $v = v(x)$.

$$\text{Let} \quad u + \delta u = u(x + \delta x)$$
$$\text{and} \quad v + \delta v = v(x + \delta x)$$

so that δu is the increase in the value of the function u as x increases by the amount δx, and δv is the corresponding increase in v.

The Derivative of a Sum $u(x) + v(x)$

$$\begin{aligned}
\frac{d}{dx}(u+v) &= \lim_{\delta x \to 0} \frac{[u(x+\delta x) + v(x+\delta x)] - [u(x) + v(x)]}{\delta x} \\
&= \lim_{\delta x \to 0} \frac{u + \delta u + v + \delta v - u - v}{\delta x} \\
&= \lim_{\delta x \to 0} \frac{u + \delta u - u + v + \delta v - v}{\delta x} \\
&= \lim_{\delta x \to 0} \frac{\delta u + \delta v}{\delta x} \\
&= \lim_{\delta x \to 0} \frac{\delta u}{\delta x} + \lim_{\delta x \to 0} \frac{\delta v}{\delta x}
\end{aligned}$$

Thus $\dfrac{d}{dx}(u+v) =. \dfrac{du}{dx} + \dfrac{dv}{dx}$

Example: The Derivative of $x^{1/2} + 3x$

$$\frac{d}{dx}(x^{1/2} + 3x) = \frac{d}{dx}x^{1/2} + \frac{d}{dx}3x = \frac{1}{2}x^{-1/2} + 3$$

The Derivative of a Product $u(x)v(x)$

$$\begin{aligned}
\frac{d}{dx}(uv) &= \lim_{\delta x \to 0} \frac{(u+\delta u)(v+\delta v) - uv}{\delta x} \\
&= \lim_{\delta x \to 0} \frac{uv + u\delta v + v\delta u + \delta u \delta v - uv}{\delta x} \\
&= \lim_{\delta x \to 0} u\frac{\delta v}{\delta x} + \lim_{\delta x \to 0} v\frac{\delta u}{\delta x} + \lim_{\delta x \to 0} \frac{\delta u \delta v}{\delta x} \\
&= u\frac{dv}{dx} + v\frac{du}{dx} + \lim_{\delta x \to 0} \frac{\delta u \delta v}{\delta x}
\end{aligned}$$

The final term may be treated as either

$$\lim_{\delta x \to 0} \delta u\frac{\delta v}{\delta x} = \delta u\frac{dv}{dx}$$

or

$$\lim_{\delta x \to 0} \delta v\frac{\delta u}{\delta x} = \delta v\frac{du}{dx}$$

In either case it will be zero, since as $\delta x \to 0$ both $\delta u \to 0$ and $\delta v \to 0$, while

$$\frac{dv}{dx} \text{ and } \frac{du}{dx}$$

will take limiting values, so that

$$\frac{d}{dx}(uv) = u\frac{dv}{dx} + v\frac{du}{dx} \tag{6.8}$$

Example: The Derivative of x^2

If we let both $u = x$ and $v = x$ we can use equation (6.8) to calculate the derivative of x^2 as follows:

$$\frac{d}{dx}(x.x) = x.1 + 1.x = 2x$$

The slopes of both $u = x$ and $v = x$ are 1; hence, their derivatives are 1. This result tells us that the slope of the curve $y = x^2$ has the value $2x$. Thus, when $x = 3$, $y = 9$ and the slope of the curve is 6.

Example: The Derivative of x^3

The same process can be used here:

$$\frac{d}{dx}(x^2.x) = x^2.1 + 2x.x = 3x^2$$

I leave it as a challenge to the student to prove that $\frac{d}{dx}x^n = nx^{n-1}$ using the method of induction. See section 5.2.

Example: The Derivative of $x\sin x$

Let $x\sin x = uv$, where $u = x$ and $v = \sin x$.

$$\frac{d}{dx}uv = u\frac{dv}{dx} + v\frac{du}{dx} \qquad \text{using equation (6.8)}$$

$$\therefore \frac{d}{dx}x\sin x = x\cos x + \sin x \times 1$$

$$= \sin x + x\cos x$$

The Derivative of a Quotient $u(x)/v(x)$

$$\frac{d}{dx}\left(\frac{u}{v}\right) = \lim_{\delta x \to 0} \frac{\frac{u+\delta u}{v+\delta v} - \frac{u}{v}}{\delta x}$$

$$= \lim_{\delta x \to 0} \frac{uv + v\delta u - uv - u\delta v}{(v + \delta v)v\delta x}$$

$$= \lim_{\delta x \to 0} \frac{v\delta u - u\delta v}{(v + \delta v)v\delta x}$$

$$= \lim_{\delta x \to 0} \frac{v\frac{\delta u}{\delta x} - u\frac{\delta v}{\delta x}}{(v + \delta v)v}$$

$$\frac{d}{dx}\left(\frac{u}{v}\right) = \frac{v\frac{du}{dx} - u\frac{dv}{dx}}{v^2} \tag{6.9}$$

Example: The Derivative of $\tan x$

$$y = \tan x = \frac{\sin x}{\cos x} = \frac{u}{v}$$

Now if $u = \sin x$ then $\dfrac{du}{dx} = \cos x$, and if $v = \cos x$ then $\dfrac{dv}{dx} = -\sin x$

$$\therefore \frac{d}{dx} \tan x = \frac{\cos x . \cos x - \sin x(-sinx)}{\cos^2 x}$$

$$\therefore \frac{d}{dx} \tan x = \frac{1}{\cos^2 x}$$

since $\sin^2 x + \cos^2 x = 1$, from section 5.6

6.8 The Chain Rule

So far we have been limited in the types of functions that we can differentiate. For example, we can differentiate the function $y = \sin x + x^2$, but we are unable to differentiate $y = \sin(x^2)$. The latter is an example of a function of a function.

Suppose that u is a function of x:

$$u = f(x)$$

and that y is a function of u:

$$y = g(u)$$

so that we may write

$$y = g(f(x))$$

Using the example mentioned already we would have

$$u = x^2$$

$$y = \sin(u) = \sin(x^2)$$

Now if we let x increase by a small amount to $x + \delta x$, we cause u to change from u to $u + \delta u$ and therefore y will change from y to $y + \delta y$. We may write the following algebraic equation relating these small increments:

$$\frac{\delta y}{\delta x} = \frac{\delta y}{\delta u} \frac{\delta u}{\delta x} \quad \text{which, if } \delta x \to 0 \text{ becomes} \quad \frac{dy}{dx} = \frac{dy}{du} \frac{du}{dx}$$

Notice that by cancelling the two du terms, the above equation is obvious!

Example: The Derivative of $\sin(x^2)$

$$
\begin{aligned}
y &= \sin(x^2) \\
\text{put } u &= x^2 \\
\text{so that } y &= \sin u \\
\text{therefore } \frac{dy}{dx} &= \frac{dy}{du}\frac{du}{dx} = \cos u \times 2x \\
\frac{d}{dx}\sin x^2 &= 2x\cos x^2
\end{aligned}
$$

I sometimes find it easier to write the function explicitly, rather than referring to an additional (abstract) function, as follows:

$$
\begin{aligned}
\frac{d}{dx}\sin x^2 &= \frac{d\sin x^2}{dx^2}\cdot\frac{dx^2}{dx} \\
&= \cos x^2\cdot 2x
\end{aligned}
$$

This is exactly the same as the above, but I prefer to leave out the "put $u = x^2$," etc., since it makes the solution longer and messy!

Example: The Derivative of $(x^2 + 3x + 1)^4$

$$
\begin{aligned}
y &= (x^2 + 3x + 1)^4 \\
\text{Let } u &= (x^2 + 3x + 1) \quad \text{so that} \quad y = u^4 \\
\text{now } \frac{dy}{du} &= 4u^3 \quad \text{and} \quad \frac{du}{dx} = 2x + 3 \\
\frac{dy}{dx} &= \frac{dy}{du}\frac{du}{dx} \\
\frac{dy}{dx} &= 4(x^2 + 3x + 1)^3(2x + 3)
\end{aligned}
$$

As in the previous example, the solution is shorter if we write it without using the abstract function:

$$
\begin{aligned}
\frac{d}{dx}(x^2 + 3x + 1)^4 &= \frac{d(x^2 + 3x + 1)^4}{d(x^2 + 3x + 1)}\cdot\frac{d(x^2 + 3x + 1)}{dx} \\
&= 4(x^2 + 3x + 1)^3\cdot(2x + 3)
\end{aligned}
$$

Example: The Derivative of e^{kt}

$$\frac{d}{dt}e^{kt} = \frac{de^{kt}}{d(kt)} \cdot \frac{d(kt)}{dt}$$
$$= e^{kt}.k$$
$$\frac{d}{dt}e^{kt} = ke^{kt} \qquad (6.10)$$

This is an important result with applications in many areas of biology, particularly in describing population growth and rates of decay such as for pollutants and trace elements. The equation states that the rate of increase or decrease of an exponential function is proportional to the function itself—the more you have, the bigger the rate of increase or decrease!

6.9 Optimum Values: Maxima and Minima

We are often asked to find the optimum (maximum or minimum) value of some function or process. For example, a manager is much more interested in the maximum profit available than in any other value of profit. We might also be asked to define values such as the minimum cost of production or the maximum growth rate.

Consider the function plotted below:

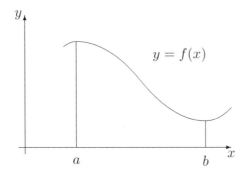

The values of y at $x = a$ and $x = b$ are maximum and minimum, respectively, because all other values in the vicinity of $x = a$ are less than $f(a)$, and all the values in the region of $x = b$ are greater than $f(b)$. The value of the function at $x = a$ and $x = b$ are not necessarily the largest or smallest values that the function can take. A function may have any number of maxima and minima.

It is perhaps stating the obvious that maxima and minima occur alternatively for any function. After a maximum value the function must decrease before increasing again to the next maximum; therefore, there must be some point between two maxima where a minimum value occurs. Similarly, there must be a maximum value between each pair of minima.

Consider the graphs in figure 6.3, where the function and its first and second derivatives are plotted. The second derivative, $d^2y/dx^2 = d/dx \cdot dy/dx$, is the slope of dy/dx or the "slope of the slope." The values of these derivatives are given in the following table.

	$\frac{dy}{dx}$	$\frac{d^2y}{dx^2}$
Maximum	0	$-ve$
Minimum	0	$+ve$
Inflection	0	0

We can use the values in the table above to identify the turning points of any specific function. Nonzero values of the second derivative allow us to identify maxima or minima, but a zero value does not necessarily identify an inflection. In order to be certain of our results, we must calculate the higher-order derivatives until we find a nonzero value. If d^ny/dx^n is the first nonzero derivative (other than dy/dx), then the turning point is an inflection if n is odd. Otherwise it will be a maximum or minimum depending upon whether d^ny/dx^n is negative or positive.

Application: How Fast Should a Fish Swim?

If the rate of energy utilization by a swimming fish is proportional to the cube of its speed, show that the most economical speed for the fish to swim against the current will be $1\frac{1}{2}$ times that of the current.

Let the speed of the current be R and the speed of the fish relative to the water be S. The speed of the fish relative to the bank, B, is given by

$$B = S - R$$

Over a given distance, x, the time taken will be

$$t = \frac{x}{B} = \frac{x}{S - R}$$

and since the rate of energy utilization is kS^3, where k is a constant, the total energy used (E) in covering the distance x is given by

$$E = kS^3 \frac{x}{S - R} = kx \frac{S^3}{S - R}$$

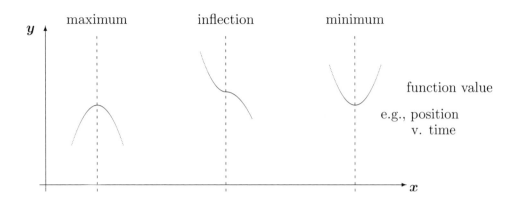

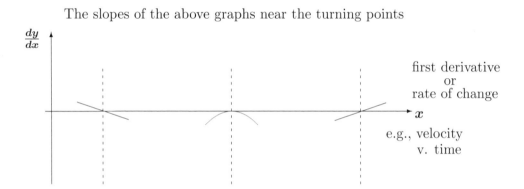

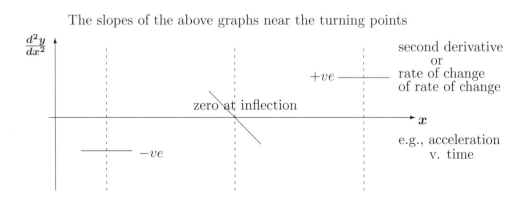

Figure 6.3: Turning Points and Their First Two Derivatives

The minimum value of E is obtained when $dE/dS = 0$. That is,

$$\frac{d}{dS}\frac{S^3}{S-R} = 0$$

$$\frac{(S-R)3S^2 - S^3}{(S-R)^2} = 0$$

$$\therefore \quad (3(S-R) - S) \times S^2 = 0$$

$$\therefore \quad 3(S-R) - S = 0$$

$$2S = 3R$$

$$\therefore \quad S = \frac{3R}{2}$$

Note the zeros at $S = 0$ are of no interest to us because the fish is not swimming!

6.10   Small Errors

From the definition of the differential coefficient

$$\frac{dy}{dx} = \underset{\delta x \to 0}{Lim}\ \frac{\delta y}{\delta x}$$

we can see that, provided δx is small,

$$\frac{\delta y}{\delta x} \approx \frac{dy}{dx}$$

and so by multiplying both sides by δx we have

$$\delta y \approx \frac{dy}{dx}\delta x$$

Inevitably there are errors in measurement during experimental processes, and the use of the differential coefficient as shown above allows us to calculate the error on derived variables. This can be an important part of the analysis when comparing errors of measurement with those associated with different treatments. See the calculation below.

Example: Estimation of Errors

In a water droplet experiment we need to calculate the volume of the droplet using a microscope to measure the diameter D (mm). If we can measure the diameter to an accuracy of $\pm\ 0.001$ mm, what is the accuracy of our calculated volume, assuming that the droplets are spherical?

$$V = \frac{4}{3}\pi\left(\frac{D}{2}\right)^3$$

$$= \frac{\pi}{6}D^3$$

$$\frac{dV}{dD} = \frac{3\pi D^2}{6} = \frac{\pi D^2}{2}$$

$$\therefore \quad \frac{\delta V}{\delta D} \approx \frac{\pi D^2}{2}$$

$$\delta V \approx \frac{\pi D^2}{2}\delta D$$

$$\approx \pm 0.0005\pi D^2 \text{mm}^3$$

6.11 Summary Notes on Differentiation

6.11.1 Standard Derivatives

$$\frac{d}{dx}f(x) = \lim_{\delta x \to 0}\frac{f(x+\delta x)-f(x)}{\delta x}$$

$$\frac{d}{dx}\ln x = \frac{1}{x}$$

$$\frac{d}{dx}\text{constant} = 0$$

$$\frac{d}{dx}x^n = nx^{n-1}$$

$$\frac{d}{dx}\sin x = \cos x$$

$$\frac{d}{dx}\cos x = -\sin x$$

$$\frac{d}{dx}e^x = e^x$$

6.11.2 Rules for Differentiation

If u and v are both functions of x:

$$\frac{d}{dx}(u+v) = \frac{du}{dx} + \frac{dv}{dx}$$

$$\frac{d}{dx}(u \cdot v) = u\frac{dv}{dx} + v\frac{du}{dx}$$

$$\frac{d}{dx}\left(\frac{u}{v}\right) = \frac{v\frac{du}{dx} - u\frac{dv}{dx}}{v^2}$$

If y is a function of θ and θ is a function of x,

$$\frac{dy}{dx} = \frac{dy}{d\theta}\frac{d\theta}{dx}$$

6.11.3   Maxima and Minima

At a *maximum* or *minimum* value of y, $dy/dx = 0$. At a *maximum* d^2y/dx^2 is *negative*, at a *minimum* it is *positive*. More generally, if $dy/dx = 0$ and the first non-zero derivative is odd (d^3y/dx^3, d^5y/dx^5, etc.), there is an inflection. However, if it is even, the turning point is a maximum or a minimum depending upon whether the sign of this derivative is negative or positive.

6.12   Applications

Equation for Radioactive Decay

Radioactive substances are used extensively to trace various chemical and biological reactions in living organisms. Accidents such as those at Chernobyl and Three Mile Island also provide ample opportunity to examine the effects of radioactive pollution. Here, we derive the fundamental equation for exponential decay.

Let $R(t)$ Bq be the radiation at time t and R_0 Bq the initial radiation.[1] The rate of increase (dR/dt) of radioactive material will be negative, and proportional to the amount of radioactive substance present:

$$\frac{dR}{dt} \propto -R \qquad \text{or} \qquad \frac{dR}{dt} = kR$$

where k is a negative constant, the *decay constant*, specific to the material.

From our experience, or by reference to the earlier example in equation (6.10), we can remember that

$$\frac{d}{dt}e^{kt} = ke^{kt}$$

where k is a constant, so that if $R = e^{kt}$ and k is negative

$$\frac{dR}{dt} = kR$$

Thus the radiation at time t may be described by the function

$$R = R_0 e^{kt} \tag{6.11}$$

where k is the decay constant and R_0 is the original level of radioactivity.

---

[1] Bq: 1 becquerel = 1 disintegration per second

Half-Life

The half-life of a radioactive substance is defined as the time taken for the level of radiation to fall by one-half. Thus, if t_{half} is the half-life, we can use equation (6.11) to give

$$\frac{R_0 e^{kt_{half}}}{R_0} = 1/2$$

$$\therefore e^{kt_{half}} = 1/2$$

$$kt_{half} = \ln(1/2)$$

$$t_{half} = \frac{\ln(1/2)}{k}$$

$$= -0.693/k$$

The half-life of polonium 210 is 138 days. What is its decay constant?

$$k = \ln(1/2)/t_{half} = -0.693/138 = -0.005$$

Fitting the Best Line: The Method of Least Squares

Here, for completeness we derive the equations used in section 4.10.2 for fitting the best straight line through a set of data points.

Suppose we have the following set of experimental values

$$(x_1, y_1), (x_2, y_2), \ldots (x_i, y_i) \ldots (x_n, y_n)$$

and we suspect that the two variables are linearly related $(y = a + bx)$.

Draw the line $y = a + bx$ as shown (at present we don't know the values of a and b—we just guess them).

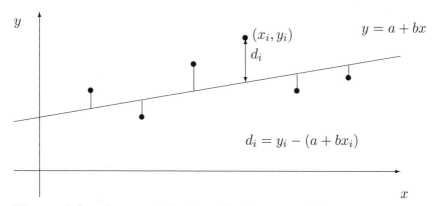

We can define the quantities d_i as the distance of the experimental points from the line (in the y direction)—i.e., the difference between the experimental value of y (y_i) and the value of y predicted by the straight line ($a + bx_i$).

$$d_i = y_i - (a + bx_i)$$

A measure of the goodness of fit can be obtained by adding the squares of all such d_i.

$$\therefore G = \sum_{i=1}^{n} d_i{}^2$$

We use the squares of the differences so that the contribution from each difference will be positive and therefore G will always be positive. If G is large, the fit is bad; if G is small, the fit is good.

$$
\begin{aligned}
G &= \sum_{i=1}^{n} [y_i - (a + bx_i)]^2 \\
&= \sum_{i=1}^{n} [y_i - a - bx_i]^2 \\
&= \sum_{i=1}^{n} [y_i^2 + a^2 + b^2 x_i^2 - 2y_i a - 2y_i bx_i + 2abx_i] \\
&= \sum_{i=1}^{n} y_i^2 + na^2 + b^2 \sum x_i^2 - 2a \sum y_i - 2b \sum x_i y_i + 2ab \sum x_i
\end{aligned}
$$

Note the term na^2, which is the result of adding together n copies of a^2.

We can minimize G in order to obtain the best fit (1) by changing a, keeping b constant (moving the line up or down), and (2) by changing b, keeping a constant (rotating the line).

$$
\begin{aligned}
\frac{\partial G}{\partial a} &= 2na - 2\Sigma y + 2b\Sigma x \\
\text{at minimum } na - \Sigma y + b\Sigma x &= 0 \quad\quad (6.12) \\
\frac{\partial G}{\partial b} &= 2b\Sigma x^2 - 2\Sigma xy + 2a\Sigma x \\
\text{at minimum } b\Sigma x^2 - \Sigma xy + a\Sigma x &= 0 \quad\quad (6.13) \\
\text{from (6.12)} \quad a &= \frac{\Sigma y - b\Sigma x}{n} \quad\quad (6.14)
\end{aligned}
$$

Substitute a from (6.14) into (6.13)

$$
\begin{aligned}
b\Sigma x^2 - \Sigma xy + \frac{(\Sigma y - b\Sigma x)\Sigma x}{n} &= 0 \\
nb\Sigma x^2 - n\Sigma xy + \Sigma y\Sigma x - b\Sigma x\Sigma x &= 0 \\
b(n\Sigma x^2 - \Sigma x\Sigma x) &= n\Sigma xy - \Sigma x\Sigma y \\
b &= \frac{n\Sigma xy - \Sigma x\Sigma y}{n\Sigma x^2 - (\Sigma x)^2} \quad\quad (6.15)
\end{aligned}
$$

Thus we calculate b from equation (6.15) and then use its value in equation (6.14) to calculate a.

Cylinder of Minimum Surface Area

A hot water tank consists of a closed cylinder of height H and radius R. In order to minimize heat loss, and hence save the planet and our heating bill, we would like the tank to have minimum surface area. If the volume is fixed, what ratio of H to R will give the minimum total surface area? (The same principles apply to microbial cells or cylindrical cows!)

$$V = \pi R^2 H = \text{constant} \tag{6.16}$$
$$A = 2\pi RH + 2\pi R^2 \tag{6.17}$$

From equation (6.16)

$$H = \frac{V}{\pi R^2}$$

Substitute in equation (6.17)

$$A = 2\pi R \frac{V}{\pi R^2} + 2\pi R^2$$
$$= \frac{2V}{R} + 2\pi R^2$$
$$\frac{dA}{dR} = -\frac{2V}{R^2} + 4\pi R$$

At minimum $dA/dR = 0$; hence

$$4\pi R = \frac{2V}{R^2}$$

Replace V from equation (6.16) to give

$$4\pi R = \frac{2\pi R^2 H}{R^2}$$

and hence for minimum area

$$\frac{H}{R} = 2$$

6.13   Exercises

1. Find the following derivatives.

   (a) $\frac{d}{dt} \sin t$

   (b) $\frac{d}{dt} \cos^2 t$

   (c) $\frac{d}{dt} \sin(t^2)$

   (d) $\frac{d}{dz} 3ze^z$

   (e) $\frac{d}{dx} \ln(3x^2)$

2. The milk yield of the cows in a herd has been found to follow a curve
   of the type

$$y = Ate^{-Bt}$$

   where y is the yield in liters/day and t is the time in days from the
   start of lactation. A specific animal was recorded as giving 15.9 liters
   on day 10 and 24.7 liters on day 20. What will be the milk yield for
   this animal at day 100?

   When will the maximum yield for this cow occur and what will it be?

3. (a) *From first principles* prove that $\frac{d}{dx} x^2 = 2x$

   (b) A crop is observed to be infected with a form of rust. A survey
       shows that the number of plants infected at present is three per
       square meter. The disease is known to spread according to the
       equation

$$\frac{dN}{dt} = 2\sqrt{N}$$

   where N is the number of infected plants per square meter and t
   is the time in days. When the infected rate reaches $100/\text{m}^2$, it is
   felt that the crop cannot be saved. How many days do we have
   in which to effect a cure? (Hint: The two parts of the question
   are related.)

4. Find the derivative of each of the following

   (a) $x^{3/2}$

   (b) $x^2 e^{3x}$

   (c) $\sin 3x$

   (d) $\ln 2x$

   (e) $\tan x (= \frac{\sin x}{\cos x})$

   (f) $e^{x \sin x}$

5. Find the following derivatives

    (a) $\frac{d}{dx}(a + bx + cx^2)$

    (b) $\frac{d}{dt}(\sin wt)$

    (c) $\frac{d}{dx}\left(\frac{\sin x}{x}\right)$

    (d) $\frac{d}{d\sin x}(\sin^2 x)$

    (e) $\frac{d}{dx}(3x^2 + x)\sin(2x)$

    (f) $\frac{d}{dx}e^{\cos(\ln(x))}$

6. The specific weight of water at $t°C$ is given by

$$w = 1 + (5.3 \times 10^{-5})t - (6.53 \times 10^{-6})t^2 + (1.4 \times 10^{-8})t^3$$

    Find the temperature at which water has maximum specific weight.

7. An animal is fed on a diet in which the concentration, F of an added factor is varied. It is found that the daily consumption, D, of the diet is related to the concentration of added factor as follows:

$$D = A - aF$$

    where A is the consumption when no factor is added and a is a constant. What value of F should be chosen so that the animal consumes the maximum amount of the factor?

6.14   Answers

1. (a) $\cos t$

   (b) $-2\sin t \cos t$

   (c) $2t\cos t^2$

   (d) $3ze^z + 3e^z = 3(z+1)e^z$

   (e) $2/x$

2. $A = 2.04704$, $B = 0.02527$, 16.36 liters, 29.8 liters on day 39

3. (b) 8.27 days

4. (a) $(3/2)x^{1/2}$

   (b) $e^{3x}(2x + 3x^2)$

   (c) $3\cos 3x$

   (d) $1/x$

   (e) $1/\cos^2 x$

   (f) $(sinx + x\cos x)e^{x\sin x}$

5. (a) $b + 2cx$

   (b) $w\cos wt$

   (c) $(\cos x/x) - (\sin x/x^2)$

   (d) $2\sin x$

   (e) $2(3x^2 + x)\cos(2x) + (6x + 1)\sin(2x)$

   (f) $-(1/x)\sin(\ln(x))e^{\cos(\ln(x))}$

6. 4.113°C

7. $A/2a$

Chapter 7

Integral Calculus

7.1 Introduction

Integration (anti-differentiation) is the inverse of differentiation. We begin with an expression defining the rate of change (growth rate, heat flux, rate of infection, glucose in and glucose out, ...) and use it to define the state of the system (size, temperature, magnitude of the epidemic, amount of glucose in pool, ...).

Assume two functions $y(x)$ and $g(x)$ so that

$$\frac{d}{dx} y(x) = g(x)$$

In other words, the derivative of $y(x)$ is $g(x)$. Then $y(x)$ is the *integral* of $g(x)$, which we write as follows:

$$\text{if } \frac{d}{dx} y(x) = g(x) \text{ then } \int g\,dx = y + c \tag{7.1}$$

Here c is an arbitrary constant, which must be included because the derivative of a constant is zero. Equation (7.1) is the basis of integral calculus. An integral such as this, in which there is an unknown constant, is called an *indefinite integral*. The value of c can usually be found because there will be a boundary condition at which the values of both x and y will be known so that substituting these values will allow the calculation of c. Here are some examples that you have seen in the previous chapter:

$$\frac{d}{dx} x^2 = 2x \quad \therefore \quad \int 2x\,dx = x^2 + c$$

$$\frac{d}{dx} e^x = e^x \quad \therefore \quad \int e^x\,dx = e^x + c$$

$$\frac{d}{dx} \sin x = \cos x \quad \therefore \quad \int \cos x\,dx = \sin x + c$$

where c is an unknown constant.

If at this point you say to yourself, "So what?" I for one have sympathy with you. All we have gained so far is another definition, but we don't understand what it is all about. Accepting things because somebody says so is hardly science, so let us try to see why and how integration works.

First of all let us see if we can make sense of the notation

$$\int y\, dx$$

which is read as "the integral ($\int$) of the function (y) with respect to x (dx)." Here, x is an independent variable upon which the variable y depends. The above expression represents the sum of all the values of $y\, dx$ over the range of x. For example, if $y = 1$, $\int 1\, dx = \int dx = x$, which is the sum of all the infinitely small bits of x, which of course will be equal to x. Similarly, $\int dy = y$ or $\int d\,anything = anything$.

7.2   Integration as the Area under a Curve

Consider a curve whose equation, $y = f(x)$, is known.

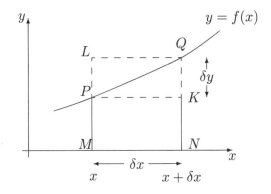

In particular, examine a small section of the curve between x and $x + \delta x$. We see that the area δA defined by the x-axis, the ordinates at x and $x+\delta x$, and the curve $y = f(x)$ lies between the areas of the rectangles $MLQN$ and $MPKN$:

$$\text{Area } (MPKN) \;<\; \delta A \;<\; \text{Area } (MLQN)$$
$$y \cdot \delta x \;<\; \delta A \;<\; (y + \delta y) \cdot \delta x$$
$$\therefore \quad y \;<\; \frac{\delta A}{\delta x} \;<\; y + \delta y$$

$$\text{Now as } \delta x \to 0 \quad \frac{\delta A}{\delta x} \to \frac{dA}{dx} = y$$

But from our definition of integration [eq. (7.1)]

$$\text{if } \frac{dA}{dx} = y$$

$$A(x) = \int y \, dx \tag{7.2}$$

That is, the area under the curve is given by the integral of the function $y = f(x)$.

However, there is still a problem with this interpretation of integration, since the area calculated using equation (7.2) will always contain an arbitrary constant, which implies that the area can have any value!

This is not such a difficulty when we think about it more carefully, because in order to define an area, we must define a *closed* area. It is impossible to calculate the area defined by the x-axis and $y = f(x)$ if we don't also define the left and right boundaries. Thus, when we are using integration to calculate the area under a curve, we must define the boundaries. We write this definition as follows:

$$area = \int_a^b y \, dx \tag{7.3}$$

where a and b are the lower and upper boundaries of the area to be found. The area under the curve is then calculated as

$$\begin{aligned} area &= (A(b) + \text{constant}) - (A(a) + \text{constant}) \\ &= A(b) - A(a) \end{aligned}$$

since the two values of the constant cancel out.

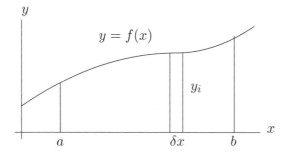

We can think of equation (7.3) as the sum of the areas of all strips of width δx under the curve between $x = a$ and $x = b$, where the area is made up of an infinite number of extremely narrow strips. The integral sign, $\int$, is derived from a long S, a shorthand notation for "sum."

$$\text{area} \quad = \quad \underset{n\to\infty}{Lim} \sum_{i=1}^{n} y_i \cdot \delta x$$

$$\text{where } \delta x = \frac{b-a}{n} \text{ and } y_i \text{ is the height of strip } i$$

$$= \quad \int_a^b y\,dx = [A(x)]_a^b \equiv A(b) - A(a) \tag{7.4}$$

The integral depicted above and in equation (7.3) is known as a *definite integral* because it does not contain an unknown constant. Take note of the *shorthand notation* in equation (7.4)

Example: Area under the Curve $y = x$
Find the area defined by the curve $y = x$, the x axis, and the ordinates $x = 1$ and $x = 3$.

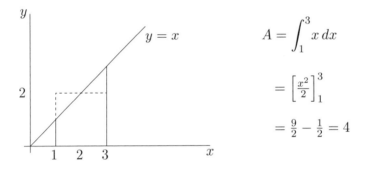

$$A = \int_1^3 x\,dx$$

$$= \left[\frac{x^2}{2}\right]_1^3$$

$$= \frac{9}{2} - \frac{1}{2} = 4$$

This is easily checked by evaluating the area directly.

Example: The Area Bound by $y = \cos x$ for $\frac{\pi}{2} \le x \le \pi$
The following is a plot of the function $y = \cos x$. A simple plot is often helpful in order to see what is going on.

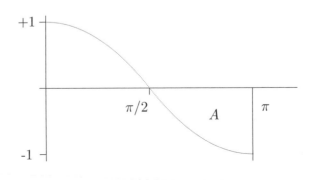

$$A = \int_{\pi/2}^{\pi} \cos x \, dx$$
$$= [\sin x]_{\pi/2}^{\pi}$$
$$= 0 - 1$$
$$= -1$$

Note that this "area" turns out to be negative! Treating the integral as an area is strictly incorrect, since the integral of a negative function gives a negative area. If we really need to calculate the area, we must integrate the negative and positive regions of the function separately and deal with the negative integrals appropriately.

Example: Evaluate the Integral $\int_0^\infty e^{-x} \, dx$

This is equivalent to finding the area under the curve e^{-x} between $x = 0$ and $x = \infty$.

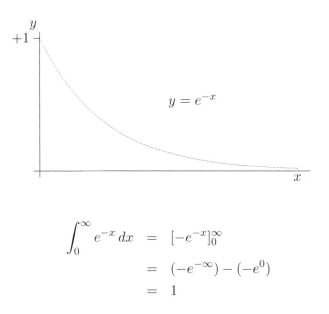

$$\int_0^\infty e^{-x} \, dx = [-e^{-x}]_0^\infty$$
$$= (-e^{-\infty}) - (-e^0)$$
$$= 1$$

Note that this area would be very difficult to calculate by direct measurement!

7.2.1 Area of a Circle 1

The concept of summing fundamental elements of area can be extended further. Consider the diagram below, which shows a small segment of a circle of radius r, bounded by radii at θ radians and $\theta + \delta\theta$ radians.

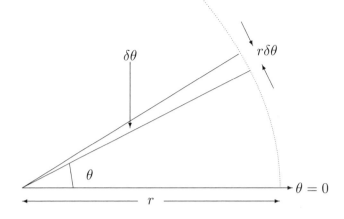

The area of such a segment, provided that $\delta\theta$ is small, and we are going to let it be infinitesimally small, can be approximated by

$$\text{area of segment} = \frac{r \times r\delta\theta}{2} = \frac{r^2\delta\theta}{2}$$

since the small segment is approximately a triangle with sides r and $r\delta\theta$. The angles at the circumference will be approximately right angles, provided $\delta\theta$ is small.

Now, we can calculate the area of the circle by integrating this expression (summing all the area segments within the circle) as follows:

$$\text{area of circle} = \int_0^{2\pi} \frac{r^2 \, d\theta}{2}$$

and because $r^2/2$ is constant, we can move it outside the integral sign to give

$$
\begin{aligned}
\text{area of circle} \;&=\; \frac{r^2}{2} \int_0^{2\pi} d\theta \\
&=\; \frac{r^2}{2} \left[\theta\right]_0^{2\pi} \\
&=\; \frac{r^2}{2}(2\pi - 0) \\
&=\; \pi r^2
\end{aligned}
$$

7.2.2   Area of a Circle 2

In a similar way we can think of a circle of radius r to be made up of many rings defined by circles of radius x and $x + \delta x$. The area of such a ring can be approximated by $2\pi x \delta x$. If we now integrate the areas of all such rings, letting x take all the values between 0 and r and $\delta x \to 0$, we have:

$$\text{area of circle} = \int_0^r 2\pi x \, dx$$

$$= 2\pi \int_0^r x \, dx$$

$$= 2\pi \left[\frac{x^2}{2} \right]_0^r$$

$$= 2\pi (r^2/2 - 0)$$

$$= \pi r^2$$

Once again we see that there is more than one way to generate a solution. It takes a little imagination (cunning!), but when we reach the same solution via two different methods, it gives us confidence in the result.

7.3 Techniques of Integration

7.3.1 The Chain Rule

The chain rule for the derivative of a function of a function is given by

$$\frac{dy}{dx} = \frac{dy}{du}\frac{du}{dx}$$
$$where \quad y = f(u) \ and \ u = g(x)$$

The corresponding formula for integration is given by

$$\int y \, dx = \int y \frac{dx}{du} \, du$$

Here we treat the function y as a function $y = f(u)$ where $u = g(x)$—that is we make a substitution which will simplify the integral.

Example: $\int \sin^2 x \cos x \, dx$

$$\text{If we put} \quad u = \sin x$$
$$\frac{du}{dx} = \cos x \quad \therefore \quad \frac{dx}{du} = \frac{1}{\frac{du}{dx}} = \frac{1}{\cos x}$$
$$\text{then} \quad \int \sin^2 x \cos x \, dx = \int u^2 \cos x \frac{1}{\cos x} \, du = \int u^2 \, du$$

$$= \frac{u^3}{3} + c$$

$$= \frac{1}{3} \sin^3 x + c$$

As with using the chain rule for differentiation I sometimes find it easier to omit the substitution and proceed as follows:

$$\int \sin^2 x \cos x \, dx$$

$$= \int \sin^2 x \, d\sin x \qquad \text{because } d\sin x / dx = \cos x \text{ and hence } \cos x \, dx = d\sin x$$

$$= \frac{1}{3} \sin^3 x + c$$

Example: $\int_0^2 e^{-x^2} x \, dx$

In order to simplify the equation make the substitution

$$u = -x^2 \quad \text{so that} \quad \frac{du}{dx} = -2x \quad \text{and } dx = \frac{du}{-2x}$$

$$\text{hence } I = \int_0^{-4} e^u x \frac{1}{-2x} \, du$$

The limits of the integration have changed, since we are now integrating over the scale of u—i.e., from 0 to $-(2^2)$.

$$I = -\frac{1}{2} \int_0^{-4} e^u \, du$$

$$= -\frac{1}{2} e^u \Big|_0^{-4}$$

$$= -\frac{1}{2} e^{-4} + \frac{1}{2} e^0$$

$$= \frac{1}{2}(1 - e^{-4})$$

Again, it sometimes simplifies the solution if we leave out the overt substitution step:

$$I = \int_0^2 e^{-x^2} x \, dx$$

$$= \int_0^4 e^{-x^2} \frac{dx^2}{2} = \frac{1}{2} \int_0^4 e^{-x^2} \, dx^2$$

$$= \frac{1}{2} \left[-e^{-x^2} \right]_0^4 = \frac{1}{2} \left[e^0 - e^{-4} \right]$$

$$= \frac{1}{2} \left[1 - e^{-4} \right]$$

7.3.2 Integration by Parts

The differential of a product is defined in equation (6.8), reproduced below.

$$\frac{d}{dx} u.v = u\frac{dv}{dx} + v\frac{du}{dx} \tag{7.5}$$

$$e.g. \frac{d}{dx} x \sin x = x\frac{d}{dx} \sin x + \sin x \frac{d}{dx} x$$

$$= x\cos x + \sin x$$

We can integrate equation (7.5) as follows

$$\int \frac{d}{dx}(u.v)\,dx = \int u\frac{dv}{dx}\,dx + \int v\frac{du}{dx}\,dx$$

to give

$$uv = \int u\,dv + \int v\,du$$

$$\text{or} \int u\,dv = uv - \int v\,du \tag{7.6}$$

The splitting of an integral suggested by the above formula will sometimes help to reduce the complexity of difficult integrals. It requires practice to see that this method will help when a substitution will not. There is a measure of artistry in the solution of integrals!

Example: $\int xe^x\,dx$

$$\text{Put } u = x \qquad dv = e^x\,dx$$

$$\frac{du}{dx} = 1 \qquad \frac{dv}{dx} = e^x$$

$$du = dx \qquad v = e^x$$

$$\int x\,e^x\,dx = x\,e^x - \int e^x\,dx$$

$$= xe^x - e^x + c$$

$$= e^x(x-1) + c$$

Example: $\int x^2 \sin x \, dx$

$$\begin{aligned}
\text{Put } u &= x^2 \\
\text{so that } \frac{du}{dx} &= 2x \\
\text{and hence } du &= 2x \, dx
\end{aligned}$$

$$\begin{aligned}
\text{and put } dv &= \sin x \, dx \\
\text{so that } \frac{dv}{dx} &= \sin x \\
\text{and hence } v &= -\cos x
\end{aligned}$$

Now substituting the above in equation (7.6) we have

$$\int x^2 \sin x \, dx = -x^2 \cos x + 2 \int x \cos x \, dx$$

Note that this integral involves only x^1 as opposed to x^2; hence, things are improving, and we should proceed along the same track.

$$\begin{aligned}
\text{Now put } u &= x & \text{so that } du = dx \\
\text{and } dv &= \cos x \, dx & \text{so that } v = \sin x
\end{aligned}$$

Substituting the above in equation (7.6) gives

$$\begin{aligned}
\int x \cos x \, dx &= x \sin x - \int \sin x \, dx \\
&= x \sin x + \cos x + c \\
\therefore \int x^2 \sin x \, dx &= -x^2 \cos x + 2x \sin x + 2 \cos x + c' \\
&= \cos x (2 - x^2) + 2x \sin x + c'
\end{aligned}$$

Example: The Factorial (!) Function

The factorial function

$$\int_0^\infty x^n e^{-x} \, dx$$

has important applications in statistics. You will come across it when using the Poisson and binomial distributions. You have already seen it in the context of the exponential function. When n is an integer, "factorial n" can be expressed as follows:

$$n! = n \times (n-1) \times (n-2) \cdots \times 2 \times 1$$

and its value represents the number of different ways that n distinct objects can be arranged in order.

$$\text{Show that} \quad \int_0^\infty x^n e^{-x}\, dx \;=\; n!$$

$$\int_0^\infty x^n e^{-x}\, dx \;=\; [-x^n e^{-x}]_0^\infty + \int_0^\infty n x^{n-1} e^{-x}\, dx$$

$$= \; 0 + n \int_0^\infty x^{n-1} e^{-x}\, dx$$

$$= \; n(n-1) \int x^{n-2} e^{-x}\, dx$$

$$\vdots$$

$$= \; n(n-1)\cdots 1 \int x e^{-x}\, dx$$

$$\int_0^\infty x e^{-x}\, dx \;=\; -x[e^{-x}]_0^\infty + \int_0^\infty e^{-x}\, dx$$

$$= \; 0 - [e^{-x}]_0^\infty$$

$$= \; -e^{-\infty} + e^0 = 1$$

$$\therefore \int_0^\infty x^n e^{-x}\, dx \;=\; n!$$

7.4 Summary Notes on Integration

7.4.1 Standard Integrals

$$\int x^n\, dx \;=\; n x^{n+1} + c \qquad \text{provided } n \neq -1$$

$$\int x^{-1}\, dx \;=\; \ln x + c$$

$$\int \sin x\, dx \;=\; -\cos x + c$$

$$\int \cos x\, dx \;=\; \sin x + c$$

$$\int e^x\, dx \;=\; e^x + c$$

where c is an unknown constant.

7.4.2   Techniques

If $\int f(x)\,dx = g(x)$, then

$$\int f(ax)\,dx = \frac{1}{a}\,g(x)$$

If $y = f(u)$ and $u = g(x)$, then

$$\int y\,dx = \int y\frac{dx}{du}\,du$$

If both u and v are functions of x, then

$$\int u\,dv = uv - \int v\,du$$

7.5   Applications

Mean Value

One very useful result of integration is

$$\frac{\int_a^b f(x)\,dx}{b-a} = \text{ mean value of } f(x)$$

Example: The Mean Value of $y = x$ for $a \le x \le b$

If $y = f(x) = x$, then the mean value of y over the interval $a \le x \le b$ is

$$\bar{y} = \int_a^b \frac{x\,dx}{b-a} = \frac{1}{b-a}\int_a^b x\,dx = \frac{1}{b-a}\left[\frac{x^2}{2}\right]_a^b$$

$$= \frac{b^2 - a^2}{2(b-a)} = \frac{(b-a)(b+a)}{2(b-a)} = \frac{b+a}{2}$$

Surfaces and Volumes of Revolution

We have seen that the integral can be interpreted as the sum of an infinite number of strips in determining the area under the curve.

We can extend this idea to find the volume, surface area, etc., of surfaces of revolution. A surface of revolution is created by rotating a two-dimensional object about an axis.

As an illustration let us calculate the volume of a circular cone shown in the drawing. The cone could be generated by "revolving" a triangle about the x-axis. The cone has a base radius, R, on the right, and its height will be H, though I have drawn it on its side, and its apex is at the origin.

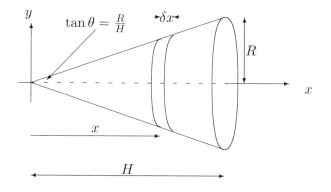

Consider a slice of thickness δx perpendicular to the x-axis at a distance x from the origin. This slice will be a disc of radius $x \tan \theta$ and thickness δx. Its volume will be approximately

$$\pi(x \tan \theta)^2 \delta x$$

Here we have ignored the fact that the edge of the disc is sloping. The error involved is of the order δx^2 (the actual term is $2\pi\theta^2\delta x^2$) and is negligible in comparison with the volume of the disc.

Now if we add together the volumes of all such discs in the range $0 < x < H$, we can calculate the volume of the cone:

$$
\begin{aligned}
V &= \int_0^H \pi(x \tan \theta)^2 \, dx = \pi \tan^2 \theta \int_0^H x^2 \, dx \\
&= \pi \tan^2 \theta \left[\frac{x^3}{3}\right]_0^H \\
&= \frac{1}{3}\pi \tan^2 \theta H^3 \\
\text{but } H \tan \theta &= R \\
\therefore \ V &= \frac{1}{3}\pi R^2 H
\end{aligned}
$$

Equations of Motion

The following are equations governing the motion of a particle undergoing uniform acceleration:

$$s = ut + \frac{1}{2}ft^2$$
$$v^2 = u^2 + 2fs$$
$$v = u + ft$$

$$\text{where} \quad s = \text{distance travelled at time } t$$
$$v = \text{velocity}$$
$$f = \text{acceleration } (= \text{constant})$$
$$u = \text{initial velocity}$$

The above equations can be derived directly from Newton's second law of dynamics. This law states that "the rate of change of momentum is proportional to the applied force, and is in the direction in which the force acts." It can be written mathematically as follows:

$$\frac{d}{dt}(mv) = P$$

where m = mass, v = velocity, and P = force, in appropriate units.

$$\frac{d}{dt}(mv) = m\frac{dv}{dt} + v\frac{dm}{dt} = P$$

In general it may be assumed that m is constant and that the equation therefore reduces to

$$\frac{d}{dt}(mv) = m\frac{dv}{dt}$$

The assumption that mass is invariant held back the development of physics for a long time until Poincaré and Einstein questioned it to produce the theory of relativity.

If we express the force as $P = mf$, where f is a constant with appropriate units, we now have

$$m\frac{dv}{dt} = mf$$
$$\therefore \quad \frac{dv}{dt} = f \qquad \text{i.e., acceleration} = \text{constant}$$

Integrating gives
$$v = ft + c$$

But if $v = u$, when $t = 0$, it follows that $c = u$.

$$\therefore \quad v \; = \; u + ft = \frac{ds}{dt} \tag{7.7}$$

$$\text{and } s \; = \; \int (u + ft)dt$$

$$= \; ut + \frac{1}{2}ft^2 + d$$

$$\text{if } s \; = \; 0 \quad \text{when} \quad t = 0 \quad \text{then} \quad d = 0$$

$$\therefore \quad s \; = \; ut + \frac{1}{2}ft^2 \tag{7.8}$$

$$\text{Also} \quad v \; = \; u + ft$$

$$\therefore \quad v^2 \; = \; u^2 + 2ftu + f^2t^2$$

$$= \; u^2 + 2f(ut + \frac{ft^2}{2})$$

substituting equation (7.8) gives

$$v^2 = u^2 + 2fs \tag{7.9}$$

Pollution of a Lake

A lake has volume $V\,(\mathrm{m}^3)$ and is fed by a river which flows through at a rate $R\,(\mathrm{m}^3/\mathrm{sec})$. If a small volume $P_0\,(\mathrm{m}^3)$ of pollutant is accidentally dropped into the lake, we would like to derive an expression for the amount of pollutant in the lake at subsequent time t, assuming that perfect mixing occurs.

We would also like to derive an expression for the time when the concentration of pollutant in the lake has fallen to one-tenth of its initial value.

To derive an expression for the amount of pollutant, let the amount of pollutant at time t be $P(t)$, so that the rate of loss of pollutant from the lake at any time t will be the concentration $(P(t)/V)$ multiplied by the outflow rate:

$$\frac{dP}{dt} = -\frac{P(t)}{V} \times R = -\frac{R}{V} \times P(t) \tag{7.10}$$

Observation that dP/dt is proportional to P should sound *exponential* warning bells. Equations of the form

$$\frac{dy}{dx} = ky \tag{7.11}$$

are satisfied by functions of the type $y = e^{kx}$, or more generally by

$$y = Ae^{kx} \tag{7.12}$$

where A and k are constants. Comparison of equation (7.10) with equations (7.11) and (7.12) leads by analogy to the following solution:

$$P(t) = Ae^{-(R/V)t} \tag{7.13}$$

However, when $t = 0$ we know that $P(0) = P_0$; hence, $A = P_0$ and we have

$$P(t) = P_0 e^{-(R/V)t} \tag{7.14}$$

The term (R/V) is often referred to as the *rate constant* in this sort of problem.

In order to find the time when the amount is reduced by a factor of 10, we require t such that

$$\frac{P(t)}{P_0} = 0.1$$

Hence,

$$
\begin{aligned}
\frac{P_0 e^{-(R/V)t}}{P_0} &= 0.1 \\
e^{-(R/V)t} &= 0.1 \\
-(R/V)t &= \ln(0.1) \\
t &= -(V/R)\ln(0.1)
\end{aligned}
$$

7.6 Exercises

1. Calculate each of the following indefinite integrals.

   (a) $\int x^2 \, dx$

   (b) $\int \cos x \, dx$

   (c) $\int (a + bx + cx^2) \, dx$

   (d) $\int \frac{1}{x} \, dx$

   (e) $\int x^2 \, dx^2$

2. A wrench was dropped from the top of a building.

   (a) Assuming that the gravitational constant g is $10 \, \text{ms}^{-1}$, its velocity, v, at time t will be given by $v = 10t$ m/sec. If it took 3 seconds to reach the ground, how high is the building?

   (b) If the wrench has mass m, its potential energy at the top of the building is given by mgh and its kinetic energy just before hitting the ground at velocity v by $mv^2/2$. Calculate both energy values and comment on your results.

3. The surface area A of a sphere of radius r is given by

$$A = 4\pi r^2$$

   (a) If a spherical cell of radius r increases its radius by δr, what will be its increase in volume?

   (b) Using the previous result, find an expression for the volume of the cell if its radius is R.

4. A radioactive source decays according to the relationship

$$C = C_0 \, e^{-kt}$$

   where C is the rate of emission at time t. C_0 is the original source strength and k is a constant. Derive an expression for R if

$$R = \int_0^T C_0 \, e^{-kt} \, dt$$

   Sketch a graph of R against T and explain the result. In particular, what is the significance of the expression C_0/k?

7.7   Answers

1. (a) $x^3/3 + c$

   (b) $\sin x + c$

   (c) $ax + bx^2/2 + cx^3/3 + d$

   (d) $\ln x + c$

   (e) $x^4/2 + c$ hint: $(x^2)^2 + c$

2. (a) 45 m

   (b) Both values are 450. The potential energy lost in falling is converted to kinetic energy.

3. (a) $4\pi r^2 \delta r$

   (b) $4\pi r^3/3$

4. The integral is

$$R = \frac{C_0}{k}(1 - e^{-kT})$$

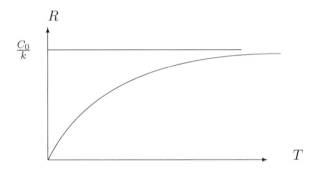

$R(T)$ represents the total radiation emitted during the period $0 - T$.

The expression C_0/k is the total radiation to be emitted. Note that the larger k, the faster the decay, and the smaller the total radiation.

Chapter 8

Matrix Algebra

8.1 Introduction

Why matrix algebra? It looks complicated, and indeed some of it can be difficult—but a little knowledge goes a long way. Could you solve 50 simultaneous equations with 50 unknowns (the nutritional properties of a diet related to its ingredients, for example)? You can if you want to, almost without effort—provided you have access to a computer. This is a case of some of the underlying mathematics being difficult, but making use of it is easy once you know how to go about it.

Many modern scientific analyses could not be achieved without the use of matrices. Therefore, it will be of use to gain a little familiarity with some of the terms. Don't be afraid; it's easier than it looks!

The objectives of this chapter are to give you sufficient knowledge to get the results, rather than to show you how it works. But it is also enlightening to see how some mathematics results from a need, rather than from a purely academic exercise.

You will be familiar with problems like the following, in which we have two equations in two unknowns.

$$3x + 4y = 18 \tag{8.1}$$
$$2x - y = 1 \tag{8.2}$$

You may also remember that in order to find a solution (the values of x and y which satisfy the equations) you need to perform some fairly tedious arithmetic. This involves reducing the equations to a single equation in one unknown and then back-substituting to find the value of the other unknown. The above is a simple example, which can be solved easily. Add $4 \times$ equation (8.2) to equation (8.1) to give

$$11x = 22$$

and hence $x = 2$. Substituting this value back into equation (8.2) produces the remaining value $y = 3$.

However, when there are more equations, writing down and solving such problems becomes much more difficult.

Mathematicians used to write sets of simultaneous equations as follows:

$$
\begin{aligned}
a_{11}x_1 + a_{12}x_2 + a_{13}x_3 + \cdots + a_{1n}x_n &= b_1 \qquad (8.3)\\
a_{21}x_1 + a_{22}x_2 + a_{23}x_3 + \cdots + a_{2n}x_n &= b_2\\
\vdots &= \vdots\\
a_{n1}x_1 + a_{n2}x_2 + a_{n3}x_3 + \cdots + a_{nn}x_n &= b_n
\end{aligned}
$$

where the a_{ij} are coefficients of a set of unknowns x_i and the b_i are values for the right-hand sides of the set of equations. Working with this notation can be very tedious, very boring, and extremely prone to errors. And so mathematicians developed a shorthand, *matrix algebra*, whereby the above equations became

$$AX = B \qquad (8.4)$$

in which A, X, and B represent complex objects.

$$
A = \begin{bmatrix}
a_{11} & a_{12} & \cdots & a_{1n}\\
a_{21} & a_{22} & \cdots & a_{2n}\\
\vdots & \vdots & & \vdots\\
a_{n1} & a_{n2} & \cdots & a_{nn}
\end{bmatrix}, \quad
X = \begin{bmatrix}
x_1\\
x_2\\
\vdots\\
x_n
\end{bmatrix}, \quad \text{and} \quad
B = \begin{bmatrix}
b_1\\
b_2\\
\vdots\\
b_n
\end{bmatrix}
$$

A is a square matrix with n rows and n columns. X and B are single-column matrices (called column vectors), each with n elements.

Equations (8.1) and (8.2) could also be represented by equation (8.4) with

$$
A = \begin{bmatrix} 3 & 4\\ 2 & -1 \end{bmatrix}, \quad
X = \begin{bmatrix} x\\ y \end{bmatrix}, \quad \text{and} \quad
B = \begin{bmatrix} 18\\ 1 \end{bmatrix}
$$

Notice that the equivalence of equations (8.3) and (8.4), and of equations (8.1, 8.2) and (8.4) define the process of matrix multiplication. See section 8.3.8.

8.2 What Is a Matrix?

A matrix is a rectangular array of numbers represented as follows:

$$A = \begin{bmatrix} a_{11} & a_{12} & \cdots & a_{1n} \\ a_{21} & a_{22} & \cdots & a_{2n} \\ \vdots & \vdots & & \vdots \\ a_{m1} & a_{m2} & \cdots & a_{mn} \end{bmatrix}$$

Matrices are normally represented by capital letters. The individual numbers a_{ij} are called elements; and when a matrix is written out in full, the elements are enclosed within square brackets. The subscripts i and j identify the row and column in which the element is located. Thus, a_{23} is the element that is located in the second row and the third column of A. It is sometimes convenient to abbreviate the matrix representation of the above to

$$A = [a_{ij}]$$

especially when $a[ij]$ is defined in terms of i and j.

A row of a matrix contains all the elements whose first subscript is the same. A column of a matrix contains all the elements whose second subscript is the same.

A matrix with m rows and n columns is called a matrix of order (m, n) or an $m \times n$ ("m by n") matrix. If $m = n$ then the matrix is said to be square and may be referred to as a matrix of order n or an n-square matrix.

8.3 Developing the Algebra

In the following notes we will develop an algebra that allows us to manipulate matrices in much the same way that we can perform arithmetic on ordinary numbers (scalars). You should note that there is nothing mysterious about this; we are simply developing a notation that will allow us to write down the steps in calculations which involve groups of similar items or equations.

8.3.1 Equality of Matrices

Two matrices $A(m, n)$, $B(m, n)$ are said to be equal if $a_{ij} = b_{ij}$ for all i, j. This means that A and B are equal only if *all* equivalent elements are equal.

8.3.2   Addition of Matrices

If $P = [p_{ij}]$ and $Q = [q_{ij}]$ are matrices of order $m \times n$, we define the sum of P and Q to be

$$
P + Q = [p_{ij} + q_{ij}]
$$

$$
= \begin{bmatrix}
p_{11} + q_{11} & p_{12} + q_{12} & \cdots & p_{1n} + q_{1n} \\
p_{21} + q_{21} & p_{22} + q_{22} & \cdots & p_{2n} + q_{2n} \\
\vdots & & & \\
p_{m1} + q_{m1} & p_{m2} + q_{m2} & \cdots & p_{mn} + q_{mn}
\end{bmatrix}
$$

Notes

$P + Q = Q + P$ is the commutative law of addition.

$P + (Q + R) = (P + Q) + R$ is the associative law of addition.

8.3.3   Subtraction of Matrices

$$
P - Q = [p_{ij} - q_{ij}]
$$

Note: We can add and subtract two matrices only if they are both of the same order.

8.3.4   Zero or Null Matrix

A matrix all of whose elements are zero is called the *zero* or *null matrix* and is usually denoted by O.

$$
A + O = A
$$

The negative of a matrix A is defined to be $-A = [-a_{ij}]$. That is, the negative of A is formed by changing the sign of every element of A. Thus,

$$
A + (-A) = O
$$

8.3.5   Transpose Matrix

The transpose of a matrix A is obtained by interchanging the rows and columns. It is denoted by A^T and is usually referred to as the transpose of A or A transpose. For instance,

$$
\text{if } A = \begin{bmatrix} 1 \\ 2 \\ 3 \end{bmatrix} \text{ then } A^T = \begin{bmatrix} 1 & 2 & 3 \end{bmatrix}
$$

$$
\text{if } B = \begin{bmatrix} 1 & 2 \\ 3 & 4 \end{bmatrix} \text{ then } B^T = \begin{bmatrix} 1 & 3 \\ 2 & 4 \end{bmatrix}
$$

8.3.6   Identity Matrix

The identity matrix is a square matrix with all its diagonal elements equal to unity and with all its other elements equal to zero.

$$I = [i_{pq}] \text{ where } i_{pq} = \begin{cases} 1 & \text{if } p = q \\ 0 & \text{if } p \neq q \end{cases}$$

The identity matrix of order 3 is

$$I = \begin{bmatrix} 1 & 0 & 0 \\ 0 & 1 & 0 \\ 0 & 0 & 1 \end{bmatrix}$$

The identity matrix has the property

$$IX = X$$

Its behavior is similar to the number 1 in ordinary algebra.

8.3.7   Multiplication by a Scalar

If k is a scalar (a number) we define $kA \equiv [ka_{ij}]$. Thus,

$$2 \begin{bmatrix} a & b \\ c & d \end{bmatrix} = \begin{bmatrix} 2a & 2b \\ 2c & 2d \end{bmatrix} = \begin{bmatrix} a & b \\ c & d \end{bmatrix} + \begin{bmatrix} a & b \\ c & d \end{bmatrix}$$

Note that every element is multiplied by the value of the scalar.

8.3.8   Matrix Multiplication

If

$$A = \begin{bmatrix} 6 & 2 & 1 \\ 3 & 1 & 2 \\ 2 & 4 & 1 \end{bmatrix} \text{ and } B = \begin{bmatrix} 3 & 6 \\ 2 & 1 \\ 1 & 2 \end{bmatrix}$$

and $C = AB$, we define c_{ij} by multiplying the elements in the ith row of A (left to right) by the corresponding elements in the jth column of B (top to bottom) and summing the products.

$$\text{Thus, } c_{11} = \begin{bmatrix} 6 & 2 & 1 \end{bmatrix} \begin{bmatrix} 3 \\ 2 \\ 1 \end{bmatrix} = 6 \times 3 + 2 \times 2 + 1 \times 1 = 23$$

$$\text{and } c_{21} = 3 \times 3 + 1 \times 2 + 2 \times 1 = 13$$

$$\text{Therefore } C = AB = \begin{bmatrix} 23 & 40 \\ 13 & 23 \\ 15 & 18 \end{bmatrix}$$

Note: AB exists only if the number of columns in A equals the number of rows in B. In the example above we say that A is postmultiplied by B and that B is premultiplied by A.

If AB exists, then A is said to be conformable to B for multiplication. The fact that A is conformable to B for multiplication does not mean that B is conformable to A for multiplication, as can be seen from the previous example.

Also, if A and B are square matrices of the same order, AB is not necessarily equal to BA, so that *matrices do not obey* the commutative law of multiplication.

8.4   Applications

Population Dynamics

The usual basis for the description of population growth is the exponential equation

$$N_t = N_0 \, e^{rt}$$

in which N_0 is the initial population and r is the "intrinsic rate of natural increase of the population." This model is limited, in that it is restricted to the situation where all individuals within the population are identical. If the model is extended (as it may be) to include different classes within the population, the mathematics becomes very complicated and the resulting model relies heavily upon the solution of integral equations.

It is possible, however, to describe a population in terms of several different age groups in a rather elegant method which makes use of matrix algebra. Here, we represent the population at a given time by a vector, the elements of which are the number of individuals in each age class.

$$\begin{bmatrix} n_1 \\ n_2 \\ n_3 \end{bmatrix}$$

In order to generate the population one time interval later, we multiply this vector by a square matrix as shown:

$$\begin{bmatrix} f_1 & f_2 & f_3 \\ p_1 & 0 & 0 \\ 0 & p_2 & 0 \end{bmatrix} \begin{bmatrix} n_1 \\ n_2 \\ n_3 \end{bmatrix} = \begin{bmatrix} f_1 n_1 + f_2 n_2 + f_3 n_3 \\ p_1 n_1 \\ p_2 n_2 \end{bmatrix}$$

This matrix is known as the Leslie matrix, and its elements correspond to the fecundities and survival rates for the individual age classes. It can be seen that elements in the top row give contributions to the youngest age group—the new value of n_1. The fecundity values (f_i) represent the number of offspring born to each individual in the relevant age group. The elements

(p_i) below the diagonal have the effect of moving members from one age group to the next higher age group; they correspond to the survival rates for each age class.

As an example consider the population described by the following Leslie matrix

$$\begin{bmatrix} 0 & 9 & 12 \\ \frac{1}{3} & 0 & 0 \\ 0 & \frac{1}{2} & 0 \end{bmatrix}$$

with an initial population containing one individual in the eldest group:

$$\begin{bmatrix} 0 \\ 0 \\ 1 \end{bmatrix}$$

Successive generations of the population are predicted by repeated multiplication as follows:

Time		0	1	2	3	4	5	6	7
Population	young	0	12	0	36	24	108	144	372
	middle	0	0	4	0	12	8	36	48
	old	1	0	0	2	0	6	4	18
	total	1	12	4	38	36	122	184	438

Initially the population structure oscillates, but it can be seen that the relative sizes of the age groups is stabilizing and that the population increase in each generation seems to be settling down to some steady value.

The ratio of numbers within each age class gradually stabilizes (in fact, the ultimate ratio is 24:4:1), and the population doubles at each time interval. This factor corresponds to an intrinsic rate of increase of 2.

It is possible to predict the rate of increase and the stable population patterns from the Leslie matrix without repeated multiplication. These properties are related to the eigenvalues and vectors of the matrix (see section 8.6.1).

Using Matrix Multiplication to Rotate Coordinates

Consider the point (1,0) in Cartesian coordinates represented as a column vector

$$\begin{bmatrix} 1 \\ 0 \end{bmatrix}$$

Pre-multiplying by the rotation matrix as follows

$$\begin{bmatrix} \cos\theta & -\sin\theta \\ \sin\theta & \cos\theta \end{bmatrix} \begin{bmatrix} 1 \\ 0 \end{bmatrix} = \begin{bmatrix} \cos\theta \\ \sin\theta \end{bmatrix}$$

has the effect of rotating the point $(1,0)$ counterclockwise about the origin by an angle θ, so that its new position is $(\cos\theta, \sin\theta)$.

If we now rotate the point again, this time by an angle ϕ

$$\begin{bmatrix} \cos\phi & -\sin\phi \\ \sin\phi & \cos\phi \end{bmatrix} \begin{bmatrix} \cos\theta \\ \sin\theta \end{bmatrix} = \begin{bmatrix} \cos\phi\cos\theta - \sin\phi\sin\theta \\ \sin\phi\cos\theta + \cos\phi\sin\theta \end{bmatrix}$$

the two rotations have the same effect as rotating by an angle $\theta+\phi$, so that the new coordinates will be

$$\begin{bmatrix} \cos(\theta+\phi) \\ \sin(\theta+\phi) \end{bmatrix} = \begin{bmatrix} \cos\phi\cos\theta - \sin\phi\sin\theta \\ \sin\phi\cos\theta + \cos\phi\sin\theta \end{bmatrix} \tag{8.5}$$

The above relationships are difficult to prove without recourse to matrix methods but are extremely useful.

Finding Pathways

Consider the diagram below, which depicts the paths through a network consisting of five compartments or nodes. Thus, there are possible pathways from 1 to 2, 2 to 3, 3 to 4, 4 to 3, and so on . Physically this could represent the flow of a drug or a metabolite through various organs within the body or of a pollutant through a water system, the possible flights between five airports, or the links in a neural network.

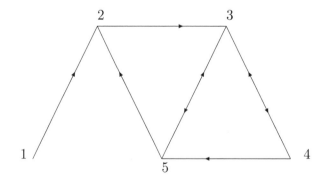

We can represent the diagram as a matrix C, known as the connectivity matrix, which shows the individual paths within the system as follows:

$$C = \begin{bmatrix} 0 & 1 & 0 & 0 & 0 \\ 0 & 0 & 1 & 0 & 0 \\ 0 & 0 & 0 & 1 & 1 \\ 0 & 0 & 1 & 0 & 1 \\ 0 & 1 & 1 & 0 & 0 \end{bmatrix}$$

The matrix shows that there is a flow (*path*) from pool 1 to pool 2, from pool 2 to pool 3, from pool 5 to pool 2, and so forth.

If we calculate $C \times C$ or C^2, the result shows the number of possible routes between the pairs of points which involve two paths.

$$C^2 = \begin{bmatrix} 0 & 0 & 1 & 0 & 0 \\ 0 & 0 & 0 & 1 & 1 \\ 0 & 1 & 2 & 0 & 1 \\ 0 & 1 & 1 & 1 & 1 \\ 0 & 0 & 1 & 1 & 1 \end{bmatrix}$$

Thus, there is a route from 1 to 2 involving two paths, a route from 2 to 4, two routes from 3 to 3 (3 to 4, 4 to 3 and 3 to 5, 5 to 3), and so on.

If we calculate C^3, the resulting elements show the number of different routes between each pair of elements involving three individual paths as follows:

$$C^3 = \begin{bmatrix} 0 & 0 & 0 & 1 & 1 \\ 0 & 1 & 2 & 0 & 1 \\ 0 & 1 & 2 & 2 & 2 \\ 0 & 1 & 3 & 1 & 2 \\ 0 & 1 & 2 & 1 & 2 \end{bmatrix}$$

The above procedure allows us to gain an insight into the pathways involved in the network and to calculate whether the pools may be linked and by which route.

8.5 Determinants

The *determinant* of a square matrix is a specific scalar—i.e., a single number—that is denoted by $\det(A)$ or $|A|$ and is closely associated with the matrix inverse. If $|A|$ is nonzero then A will have an inverse. If $|A| = 0$ the matrix does not have an inverse and is said to be singular.

The determinant of a 2×2 matrix is calculated as follows:

$$\begin{vmatrix} a & b \\ c & d \end{vmatrix} = ad - bc$$

8.5.1 The Determinant of a 3×3 Matrix

$$\begin{vmatrix} a & b & c \\ d & e & f \\ g & h & i \end{vmatrix} = a \begin{vmatrix} e & f \\ h & i \end{vmatrix} - b \begin{vmatrix} d & f \\ g & i \end{vmatrix} + c \begin{vmatrix} d & e \\ g & h \end{vmatrix}$$

$$= aei - afh - bdi + bfg + cdh - ceg$$

8.5.2   Minors and Cofactors

Let A be an n-square matrix.

If M_{ij} is the $(n-1)$-square matrix obtained by deleting the ith row and j^{th} column from A, the determinant of M is called the minor of element a_{ij}.

The cofactor of a_{ij}, denoted by A_{ij}, is given by

$$A_{ij} = (-1)^{i+j}|M_{ij}|$$

Note that A_{ij} is a scalar.

We can form a matrix of cofactors from any square matrix; for example,

$$\text{If}\quad A = \begin{bmatrix} 1 & 2 & 3 \\ 4 & 1 & 0 \\ 2 & 0 & 1 \end{bmatrix} \quad \text{the matrix of cofactors is} \quad \begin{bmatrix} 1 & -4 & -2 \\ -2 & -5 & 4 \\ -3 & 12 & -7 \end{bmatrix}$$

The determinant can be evaluated using the above notation by taking any row or column of the matrix and summing the products of each element and its cofactor.

$$|A| \quad = \quad \sum_{j=1}^{n} a_{ij} A_{ij} \qquad \text{for row } i$$

$$\text{or}$$

$$= \quad \sum_{i=1}^{n} a_{ij} A_{ij} \qquad \text{for column } j$$

The determinant of matrix A above is -13.

More generally, if B is of order 3,

$$|B| = -b_{12} \begin{vmatrix} b_{21} & b_{23} \\ b_{31} & b_{33} \end{vmatrix} + b_{22} \begin{vmatrix} b_{11} & b_{13} \\ b_{31} & b_{33} \end{vmatrix} - b_{32} \begin{vmatrix} b_{11} & b_{13} \\ b_{21} & b_{23} \end{vmatrix}$$

by expansion based on the second column.

Note: It pays to choose a row or column with many zeros.

8.5.3   Area of a Triangle

The area of a triangle whose vertices as represented in Cartesian coordinates are $(x_1, y_1), (x_2, y_2)$, and (x_3, y_3) is given by

$$\text{area} = \frac{1}{2} \begin{vmatrix} 1 & 1 & 1 \\ x_1 & x_2 & x_3 \\ y_1 & y_2 & y_3 \end{vmatrix}$$

So, the area of a triangle with coordinates $(0,0), (1,0), (0,1)$ will be

$$\text{area} = \frac{1}{2} \begin{vmatrix} 1 & 1 & 1 \\ 0 & 1 & 0 \\ 0 & 0 & 1 \end{vmatrix} = \frac{1}{2} \times \left(1 \times \begin{vmatrix} 1 & 0 \\ 0 & 1 \end{vmatrix} - 0 \times \begin{vmatrix} 1 & 1 \\ 0 & 1 \end{vmatrix} + 0 \times \begin{vmatrix} 1 & 1 \\ 1 & 0 \end{vmatrix} \right) = \frac{1}{2}$$

8.5.4 Some Properties of Determinants

1. The determinant of a matrix is equal to the sum of the products obtained by multiplying the elements of any row (or column) by their respective cofactors.

2. If two rows (or columns) of a determinant are interchanged, the value of the determinant changes sign.

3. A determinant in which all the elements of a row (or column) are zero has the value zero.

4. A determinant in which all corresponding elements in any two rows (or columns) are equal has the value zero.

5. The value of a determinant is unaltered by adding to the contents of any row (column) a constant multiple of the corresponding elements of any other row (column).

6. The value of a determinant is unaltered if rows and columns are interchanged.

8.6 The Inverse Matrix

For any scalar $x \neq 0$ there exists another scalar x^{-1} which we call the reciprocal or inverse of x, such that

$$xx^{-1} = 1$$

By analogy we might expect that for a given matrix A, we could find another matrix, which we would call A^{-1}, such that

$$A \cdot A^{-1} = I \tag{8.6}$$

We call the matrix A^{-1} (if it exists!) the inverse matrix of A. It is possible to calculate the inverse of a matrix by hand, but in general this is impractical for anything larger than a 4×4 matrix. Virtually every computer system will have access to a procedure for carrying out the calculation.

The practical use of the inverse matrix cannot be overstressed, since this, together with modern computing facilities, gives us the power to solve problems that were hitherto impossible. There are entire textbooks devoted to its calculation, but it is not necessary to know how the calculation is done, simply that it *can* be done.

However, some matrices do not have an inverse. They can usually be identified because they will have a determinant whose value is zero. Such a matrix is referred to as being singular. (The analogy with x and x^{-1} continues to hold, since the inverse of the number zero does not exist either!)

Singularity is usually due to the rows not being independent of each other—one of the rows can be formed as a combination of some of the other rows. In this case the offending row (equation) must be replaced,if possible, by a new one that is independent.

8.6.1   Solution of (Lots of) Simultaneous Equations

We can use the inverse matrix in order to find the solution of a set of simultaneous equations. For example,

$$3x + 4y = 32$$

$$2x + 3y = 23$$

may be expressed as

$$\begin{bmatrix} 3 & 4 \\ 2 & 3 \end{bmatrix} \begin{bmatrix} x \\ y \end{bmatrix} = \begin{bmatrix} 32 \\ 23 \end{bmatrix}$$

or in matrix notation as

$$AX = C$$

so that, if we can evaluate A^{-1}, we can proceed as follows:

$$\begin{array}{rcl} \text{Premultiplying by } A^{-1} \quad A^{-1}AX & = & A^{-1}C \\ IX & = & A^{-1}C \\ \therefore \quad X & = & A^{-1}C \end{array}$$

In this case the inverse matrix is

$$A^{-1} = \begin{bmatrix} 3 & -4 \\ -2 & 3 \end{bmatrix}$$

$$\therefore \begin{bmatrix} x \\ y \end{bmatrix} = \begin{bmatrix} 3 & -4 \\ -2 & 3 \end{bmatrix} \begin{bmatrix} 32 \\ 23 \end{bmatrix} = \begin{bmatrix} (3 \times 32) + (-4 \times 23) \\ (-2 \times 32) + (3 \times 23) \end{bmatrix} = \begin{bmatrix} 4 \\ 5 \end{bmatrix}$$

The solution of 100 equations with 100 unknowns is not exceptional these days. Its solution on a computer would be performed in a matter of seconds.

8.6.2   Eigenvalues and Eigenvectors

Eigenvalues and their corresponding vectors are closely linked to properties associated with the system described by the matrix equations. Thus, in the above application the vector of population size, and the fact that the population doubles at each time step, are linked to the dominant eigenvalue and its associated eigenvector.

A square matrix A is said to have an *eigenvalue* λ and corresponding *eigenvector* X if

$$AX = \lambda X \tag{8.7}$$

A matrix of order n will have n eigenvalues, scalar numbers with values that are not necessarily different, and n associatede eigenvectors. Eigenvectors are not uniquely defined, since, if X is an eigenvector, so is any scalar multiple cX. Eigenvectors are usually scaled so that $\Sigma x_i^2 = 1$. (This can be represented as $X^T X = 1$, where X is the column vector containing the individual values x_i.)

If we rearrange equation (8.7) as follows

$$(A - \lambda I)X = 0 \tag{8.8}$$

where I is the unit matrix, we have a system of homogeneous linear equations (the equations all have zero on the right-hand side). A necessary condition for the nontrivial solution of such a set of equations is that the determinant of the coefficients is zero. Thus,

$$\det(A - \lambda I) = \begin{vmatrix} a_{11} - \lambda & a_{12} & \cdots & a_{1n} \\ a_{21} & a_{22} - \lambda & \cdots & a_{2n} \\ \vdots & \vdots & & \vdots \\ a_{n1} & a_{n2} & \cdots & a_{nn} - \lambda \end{vmatrix} = 0 \tag{8.9}$$

The above is called the *characteristic* (or sometimes *secular*) equation of the matrix A and is a polynomial in λ that will have n roots, from which we can find the n eigenvalues.

Eigenvalues and vectors give access to a great many applications. For instance, the eigenvectors of a variance-covariance matrix form the basis of principal component analysis. They also show the possible stable states of a system, as in the Leslie matrix example above.

Chapter 9

Statistics

9.1 Introduction

I have mixed feelings about including a chapter on statistics in this book. My main objection is that it is such a large subject in its own right that it is difficult to give a general understanding within a few pages. Many books are already available on this subject, and if you find yourself heavily involved in statistical analysis, then you should read further. However, because statistics is an area which is likely to impinge upon most students, I will try to give a basic, but brief, understanding of its application and purpose.

Statistics sits uneasily under the general heading of mathematics. Most mathematicians feel uncomfortable with it on the grounds that the concept of probability is anything but exact and therefore is in conflict with the formal precision of mathematics.

Unfortunately, most science is not exact, which means that any conclusions reached cannot be accepted with certainty. Making subsequent decisions based upon these conclusions can be difficult and costly. Statistics provides a compromise that enables us to make those decisions with a quantitative measure of confidence.

The reputation of statistics has also been undermined by the poor use made of it by both politicians and the media. Often, individual data values are quoted out of context and judgments are made with little attempt at validation. Competent analysis, resulting in well-defined conclusions and confidence levels, is extremely important within research.

Consider the following scenarios:

1. Two researchers in the same company have discovered separate processes which they believe will improve the texture of bread. Both processes would involve similar expense in order to modify the current baking process. When asked by the production manager about the prospects of their research, "A" says that he thinks his process will improve the bread while "B" says that he is 95% confident that his

process is indeed an improvement. If you were the production manager, responsible for the expense of changing the process, of whom would you take most notice?

2. You are given the choice of two medical treatments, both of which may produce unpleasant side effects. One of the treatments affects 1 in 1000 people, while the other affects 1 in 100000. Which would you opt for?

Statistics allows us to calculate objective values for things which are not certain—in the cases above, 95%, 1/1000, and 1/100000. "Maybe" and "possibly" are not good enough to help us make important decisions. And beware the character who is "certain"!

Sometimes chance doesn't really matter ("Heads we go to the cinema tonight, tails we go to the bar"), but sometimes it does! Statistical analysis is concerned with the provision of objective answers to problems like the following.

1. The table below shows the grain weights for treated and untreated samples in a trial.

Treated	17.3	15.8	16.4	17.3	15.9	16.4	14.9	18.1	17.7
Untreated	16.9	14.3	15.1	15.4	16.2	14.8	15.1	14.6	13.9

Is there any evidence to show that the treatment is effective?

2. The weekly beer consumption (in pints) by individuals in London and provincial pubs was tabulated as follows:

London	15	17	10	28	6	14	10			
Provinces	10	17	27	15	30	6	15	33	19	20

Does this suggest a difference in drinking habits?

3. A taste panel produced the following table of rankings for eight samples each of two varieties of ice cream, Polar (P) and Glacier (G):

P   P   P   P   G   P   P   G   G   G   G   P   G   G   G   P

Is one of the varieties preferred more than the other?

The first two sets of data consist of numerical results, the first being from a continuous distribution, the second containing discrete values. In both cases we would expect there to be some underlying distribution (model) of the population data. Problems of this type are best analyzed by making use of the relevant underlying model. Statistical analyses that use an underlying distribution are known as *parametric methods*. Generally, such data are assumed to be sampled from a *normal* or *Gaussian* distribution, though you may occasionally come across data from other distributions such as the *Poisson* or *Binomial* distributions.

The third set of data is different in the sense that it consists of *ranked* items representing the order of preference. We cannot tell how much better the first choice was, nor indeed can we put a scale on any of the differences. It is therefore unlikely that we will be able to provide such an accurate analysis of these data. Nevertheless, there is still much information to be gained from an appropriate analysis of these results. The analysis of this kind of problem involves the use of *nonparametric* statistics, in which we are unable to make assumptions about the underlying distribution of the values.

Variation: The Difference between Individuals

I have often heard so-called "hard scientists" (you know who I mean!) tell us, "If you can't measure something accurately enough, then you should build a better meter or do the measurement more carefully."

The same scientists are able to measure the distance to the moon or Mars to within half a cat's whisker, thus making it difficult to argue. However, they don't usually get their feet wet or their hands dirty, nor do the moon or Mars behave in any random way. Living things do! So while the hotshots spend their time trying to solve the three-body problem[1] (most biological systems have millions) we have to make do with what we can extract from experimental results that are never repeatable. Thus, it is necessary to accept and to deal with variation.

In the case of the alternative medical treatments mentioned above, because our subject material has its own inherent variation, we know that the results of an experiment will show some differences, whatever we do to the treated group. Our problem, therefore, is to identify the variation due to the treatment as opposed to that of the inherent variation.

In order to do this we usually compare our experimental results with a sample which has not had the treatment (the *control* group). We then evaluate the probability that our experimental results came from the same population as the control group.

Calculating probability often involves the evaluation of fairly complicated statistics (numerical values such as mean and correlation coefficient), which we can then compare with tables that have been produced for us by theoretical statisticians and mathematicians. In effect, these researchers have performed similar experiments many times in order to calculate the probabilities of all possible results. From these tables we are able to predict

---

[1]The analytical solution of three objects moving under gravity. It should be simple but isn't!

the probability that our experimental results were achieved by chance irrespective of the treatment. If this probability is small (i.e., significant), we can be justified in assuming our experiment had some kind of effect.

Probability: The Predicted Likelihood of an Event

To gain some understanding of probability we will begin by examining the simple process of tossing a coin. What is the likelihood that a coin will come down heads, assuming it's got a head on one side and a tail on the other? Yes! 50% or 0.5 (which I prefer). We can reach this conclusion because there is an equal chance of a head or a tail, and the probability of either a head or a tail (i.e., any one of all the possible results) is 1. For the purist (or pedant!) I have ignored the (remote) possibility of the coin landing on its edge, or never coming down. If either of these happen, simply toss the coin again or use another—preferably not yours!

Suppose that you have just spun the coin three times and have seen three heads. Now, if we toss the coin again, what is the chance of a head? a tail? "The law of averages" tells Joe Public that we should be getting a tail by now (in order to balance things out!), so the chance of getting a tail must be bigger. Nonsense! The law of averages (if it exists at all) does not work in this way. The probability of tossing a head with a fair coin is 0.5, irrespective of how many heads or tails have appeared before. After a long sequence of tosses (think of a number and then double it) we would expect to see roughly the same number of heads as tails, but we would not expect to see head, tail, head, tail, etc.; otherwise, the results would not be random, and we would be able to predict the outcome.

Lesson 1: *Beware the law of averages.* It's not a law; it is simply an expectation. Even though we would expect to see roughly the same number of heads and tails from a long sequence of tosses, we cannot predict the order in which they will occur.

We can apply the same argument to the New York Lottery. Here the probabilities are slightly different: approximately 1 in 45 million of winning, and therefore $1 - 1/(45 \text{ million})$ of not winning.[2]

If you stand in line listening to people intent on buying their lottery tickets, you are almost certain to hear: "I've been using the same six numbers ever since the lottery started; it must be my turn soon!" They dare not

---

[2]The probability of winning the N.Y.L. (predicting 6 numbers from 59) is the combination of the probabilities of six successive events: the probability of drawing one of your six selections from the 59 balls, the probability of drawing one of your remaining 5 from 58 ... and the probability of drawing your last selection from the remaining 54. Here is the calculation:

$$\frac{6}{59} \times \frac{5}{58} \times \frac{4}{57} \times \frac{3}{56} \times \frac{2}{55} \times \frac{1}{54} = \frac{1}{45057474} \approx 0.00000002$$

miss now because, by the law of averages, their combination must come up shortly. Nonsense! The probability of any one combination coming up in any one lottery draw is always the same (1 in 45057474). End of lesson 1, and keep your dollars in your pocket.

We can extend our understanding of probability by looking at how to combine probabilities.

For example, we could consider the likelihood that two coin tosses will produce two heads. This is simply the probability that the first coin will produce a head (0.5), times the probability that the second coin will produce a head (0.5). Thus, the result is 0.25. An alternative method of evaluating this is to calculate the ratio of the number of ways in which we can obtain the required result divided by the total of all possible results (1/4).

A more interesting question is, "What is the probability of obtaining a head and a tail when both coins have been tossed?" We don't specify the order; a head and a tail is just as good as a tail and a head. Thus, two of the four possibilities meet our criterion, and therefore the probability is 0.5. This type of situation, where the result may consist of several different permutations of results, is known as a *combination.* In this case the combination is the set of results (HT, TH) that can make up the result of one head and one tail.

Unfortunately, the word *combination* is misused in many ways. For example, if you were told that the combination of a safe was 213, you would not expect to open it by using 123 or 321 or any of the other three possibilities (permutations). Thus, the lock on this type of safe should properly be called a permutation lock. Combination locks would be much easier to open!

We could look at other examples, as many textbooks do, and explain how statistics had its birth in various card games, but I don't intend to do this. I gave up playing bridge because of the endless arguments about probabilities which erupted after the games. To my mind, games should be fun, and while I enjoyed the mental athletics at the time, I couldn't subscribe to the serious chat afterwards. Much better to have a beer.

But, when it comes to performing experiments and analyses that relate to areas like drugs and food manufacturing, where serious consequences may result, we have an obligation to get it right. Thus, we need to evaluate our results and to perform relevant statistical analyses in order to apply our science as safely as possible.

And this is the answer to "Why do we use statistics?" Statistics gives us a quantitative evaluation of the situation (the probability of success) rather than a qualitative one ("I think it's a good idea"). Your hunches may be good, but they will be better backed up with a careful statistical analysis.

9.2   The Statistical Method

The following is a summary of the process to be followed in order to provide the statistical evidence in favor of a hypothesis.

1. State the *experimental hypothesis* (H1)—e.g., Whizzo fuel additive increases mileage.

2. Define the *null hypothesis* (H0) as *all possible alternatives* to the experimental hypothesis—e.g., Whizzo fuel additive either decreases mileage or makes no difference.

3. Define the critical probability value. This is usually 5%, 1%, or 0.1% but can be any relevant value, depending upon the importance of the conclusion.

4. Design the experiment or survey. This is the most important and most difficult step. For anything but the simplest experiment you should consult with a statistician in order to check that the subsequent analysis is capable of producing the results that you desire. Take several measurements of both treated and untreated objects in order to be able to separate treatment differences from natural (random) variation. Make sure that other factors do not interfere with the results— for example, use the same car, on the same road, under the same traffic and climatic conditions.

5. Carry out the experiment or survey and collect the results.

6. Calculate the probability of obtaining the result that you have just measured on the basis that the *null hypothesis* is true.

7. Reject the null hypothesis if this probability is less than the (predefined) critical value. Otherwise, reject the experimental hypothesis. There is no shame in failure, provided it has been achieved with rigor and honesty. It is as important to know that a treatment does not have an effect as it is to know that it does.

9.3   Basic Statistics

Statisticians frequently use the words *population* and *sample*, so it is important to understand what is meant by these terms. A population consists of all the individuals of a specified object, such as all the people living in New York, or all the polar bears in the Arctic, or all the ranches in Texas. Usually it will be impossible to consider all the members of a population for logistic reasons, and therefore we usually perform statistical analyses on samples from the population. A sample consists of a small but significant subset of a population, the individuals of which will usually be chosen at random and which should be representative of the population.

9.3.1 Mean

The mean of a population (μ) is calculated using the formula

$$\mu = \Sigma x / n$$

Usually, we are unable to calculate the value of μ because we don't have access to all the individuals (n) in the population, and so we have to make do with an estimate ($\bar{x}$) calculated from a sample containing n values

$$\bar{x} = \Sigma x / n$$

where n is now the sample size.

9.3.2 Variance

The variance of a population with a mean μ and containing n observations is

$$\sigma^2 = \Sigma(x - \mu)^2 / n$$

However, as stated above, we do not usually know μ and so must use an estimate, $\bar{x}$, instead. When we calculate the variance this way, we have to adjust the formula a little bit (because it gives answers which are slightly too small: there's some mathematical jiggery pokery here, but we can prove it if necessary). The best *estimate* of the variance (s^2) that we can calculate from a *sample* of size n turns out to be given by the formula

$$s^2 = \Sigma(x - \bar{x})^2 / (n - 1)$$

Note that the denominator is not n. We use $n-1$ degrees of freedom, arguing that one degree of freedom has been lost in estimating the mean. I like to think of degrees of freedom in the following way: If we have n observations but the mean is fixed, then we need to measure only $n - 1$ values because the remaining value must ensure that the overall mean is correct.

Note also that it is impossible to estimate the variance from a sample of one!

9.3.3 Standard Deviation

The standard deviation of a population (σ) is the square root of the variance (σ^2). The estimate of the standard deviation (s) is the square root of the variance estimate (s^2) and is a measure of how much the sample values vary about the sample mean.

9.3.4 Standard Error (of ...)

Usually when we deal with the standard error, we are concerned with the standard error of a derived value—the mean of a sample, or the difference between two sample means as in the t-test.

The standard error of a sample mean is given by

$$SE_{mean} = \frac{s}{\sqrt{n}}$$

where n is the sample size. Sometimes the term is used on its own to mean the standard error *of the sample*, which is the same as the standard deviation.

9.4 The Normal Frequency Distribution

The normal distribution is fundamental to much of modern statistical analysis. We need not concern ourselves with the underlying mathematical theory. Suffice it to say that many naturally occurring phenomena may be described in terms of the normal distribution. In addition, measurements which intrinsically are not *normal* may be combined by taking the mean of several values to give a close approximation to normality. Thus, the assumption, which is often made, that a population follows a normal distribution is generally a reasonable one.

The normal frequency distribution, figure 9.1, is a bell-shaped curve which is symmetrical about the mean (μ). Many values are observed close to the mean, while fewer occur as we move away from it.

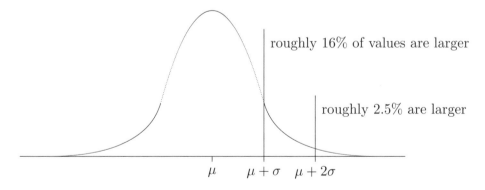

Figure 9.1: The Normal Distribution

The y-axis represents the relative frequency of observing the corresponding value of x. Note here that the area under the curve and bounded by two values of x represents the probability of observing a value between x_1 and x_2. The probability of observing a value of x between $-\infty$ and $+\infty$ (i.e., any value) is 1; therefore, the total area under the curve is 1. The probability of observing a specific value of $x(x_s,$ say) is zero—corresponding to the area bounded by the curve, the x-axis, and two overlapping lines at $x = x_s$. This has the value zero because the probability is given by the area of a strip whose width is zero.

The areas under the curve to the right of the vertical lines represent the probabilities of recording values of x greater than $x = \mu + \sigma$ and $x = \mu + 2\sigma$, respectively.

The shape of the curve will depend upon the variation of the data. If the variation is small, the curve will be sharply peaked and have a large value at the mean; if the variation is large, the curve will be more akin to a prairie with only a small value at the mean. Two properties of importance, therefore, are the variance and the mean.

The mathematical equation of the curve is

$$y = \frac{1}{\sigma\sqrt{2\pi}} e^{-(x-\mu)^2/2\sigma}$$

where μ is the mean and σ the standard deviation. The *standard normal distribution* is a normal curve with a mean of zero and standard deviation of one. The equation for the standard normal curve is therefore

$$y = \frac{1}{\sqrt{2\pi}} e^{-z^2/2} \text{ where } z = \frac{x - \mu}{\sigma}$$

The probability of obtaining values greater than z are obtained from standard normal distribution tables or computer programs. The variable z is the distance of x from the mean, in units of standard errors.

Example: Quantity Control of a Bottling Process

A bottling process is set up to dispense 501 cc of liquid, with a standard deviation of 0.5 cc, into marked 500-cc bottles. Each bottle is capable of holding 502.5 cc of liquid before overflowing.

In a batch of a 1000 filled bottles:

1. How many bottles will suffer spillage?

2. How many bottles will contain short measure?

Question 1 requires us to find the probability of dispensing quantities greater than 502.5 cc. This is the same as calculating the probability of obtaining values greater than three standard deviations above the mean $[(502.5 - 501)/0.5]$ from the standard normal curve. The computer informs me that this probability is 0.00135; hence, we should expect a spillage once in every 1000 bottles filled.

Question 2 requires us to calculate the probability of obtaining quantities less than 500 cc. This is the same as asking what the probability is of obtaining a value more than two $[(501-500)/0.5]$ standard deviations below the mean of a standard normal distribution. Since the normal distribution is symmetrical, this is equivalent to finding the probability of obtaining values greater than $+2$. This probability is 0.0227, and thus, we should expect 23 bottles in each batch to be underfilled.

Who would be interested in the answers to the above and why?

9.5   The t-Test: Are Two Means Different?

A common statistical task is the comparison of means of two samples to establish if there is a difference.

Perhaps we have a number, n_1, of patients who have been treated with a placebo, and we wish to compare them with another group comprising n_2 patients that have been given a new drug. We would like to know if the drug increases the rate of blood clotting. Data would consist of recordings of blood clotting times from patients in the two samples.

The experimental hypothesis (H1) in this case would be "The drug decreases clotting time." This leads to the null hypothesis (H0) "The drug does not decrease clotting time and therefore either increases it or leaves it the same."

The test statistic to use in this case is

$$t = \frac{\bar{x}_1 - \bar{x}_2}{\sqrt{s^2(1/n_1 + 1/n_2)}}$$

where

n_i = number of observations for sample i
$\bar{x}_i$ = estimate of the mean for sample i
s_i^2 = variance estimate for sample i
s^2 = pooled variance estimate for the population
   $= \frac{(n_1-1)s_1^2 + (n_2-1)s_2^2}{n_1+n_2-2}$

with $n_1 + n_2 - 2$ degrees of freedom.

The pooled variance estimate looks a complicated expression, but careful examination and thought will reveal that it is no more than the weighted mean of the two sample variances.

Basically, the t-statistic comprises the difference between the two means (the top line) divided by the pooled estimate of its standard error (the bottom line). This latter is a measure of the difference we would expect due to random effects if the samples were from the same population.

$$t = \frac{\text{difference between the means}}{\text{standard error of the difference}}$$

Thus, the larger the difference between the means, the larger the value of t, and the larger the standard error, the smaller the value of t. A large value of t leads us to believe that the two sample means are not the same.

You should note here that we have defined the samples so that, if H1 is correct, then the t-value should be positive. It is possible, however, that the calculated value of t could be negative. This should convince us that the drug does not improve clotting and therefore that we can dismiss the experimental hypothesis without further consideration.

The frequency distribution of the t-statistic is similar in shape to the normal distribution with a mean value of 0. However, the t-distribution for a given t-test is also dependent upon the degrees of freedom. Values showing the probability of observing values greater than t for specific values of n (degrees of freedom) are to be found in tables (or via the computer). This probability value tells us how likely it was that the two samples came from the same population

Thus a t-value of 3.365 with five degrees of freedom corresponds to a probability of 0.01—i.e., there is a chance of 1 in 100 that the mean of sample 1 could be larger than the mean of sample 2 even if H0 is true (i.e., if both samples came from the same population). An alternative and more usual interpretation of the above is to treat the result from t-tables as the probability of being wrong if we accept the experimental hypothesis. Thus, we could conclude that the drug does improve blood clotting with a 0.01 probability of being wrong. (We could also, but rather stupidly, say that the drug did not improve clotting with 0.99 probability of being wrong!)

9.6   How to Perform a t-Test

Assume that we have observed data for two samples, A and B. We would perform a t-test to verify one of three possible experimental hypotheses:

1. The average of sample A is larger than that of sample B. This is known as a *one-tailed* test, and we proceed as follows:

   (a) Calculate the means $\bar{A}$ and $\bar{B}$.

   (b) If $\bar{A} \le \bar{B}$ accept the null hypothesis, obviously! Otherwise,

   (c) Calculate the t-value $= (\bar{A} - \bar{B})/SE(\bar{A} - \bar{B})$.

   (d) Look up the probability, p, via the computer or in tables. The value of p is the probability of obtaining a larger value of t by chance when the null hypothesis is true.

   (e) If p is less than the critical value (typically 5%, 1%, or 0.1%,) accept the experimental hypothesis. Otherwise accept the null hypothesis.

2. The average of sample A is smaller than that of sample B. Reverse the sample names and proceed as above.

3. The average values of samples A and B differ. In this case we must consider both of the above hypotheses, since we are not concerned with which of the samples is larger. Hence, we must include both possibilities in our calculation. This is known as a *two-tailed* test and is carried out as follows:

   (a) Calculate t for the case that generates a positive value.

   (b) Look up the probability, p, corresponding to t, which is the probability of observing a value greater than the calculated value by chance if the null hypothesis is true. We must also take into account the possibility that the observed value is less than $-t$, which will also be p. Thus, the overall probability of observing a value which is either larger than t or smaller than $-t$ will be $2p$.

   (c) If $2p$ is less than the critical value (typically 5%, 1%, or 0.1%), accept the experimental hypothesis. Otherwise, accept the null hypothesis.

Example: Does Treatment Improve Yield?

As an example we return to a question posed at the beginning of this chapter. We are presented with data for two samples of grain weight from a trial. One of these has been treated; the other is a control. We believe that the treatment will increase the yield. The analysis goes as follows:

What type of analysis. The experimental hypothesis [H1] is "The treatment increases yield." This corresponds to a *type 1 hypothesis*. The null hypothesis is "The yield remains the same or is worse."

Calculate the parameters for both samples as in 1a.

| Treated | 17.3 | 15.8 | 16.4 | 17.3 | 15.9 | 16.4 | 14.9 | 18.1 | 17.7 |
| Untreated | 16.9 | 14.3 | 15.1 | 15.4 | 16.2 | 14.8 | 15.1 | 14.6 | 13.9 |

$$n_1 = 9 \qquad \bar{x}_1 = 16.644 \qquad s_1^2 = 1.065$$

$$n_2 = 9 \qquad \bar{x}_2 = 15.144 \qquad s_2^2 = 0.868$$

First look, as in 1b. Is the mean for sample 1 larger than the mean for sample 2? If not, the conclusion must be that H0 is correct. However, in this case the situation looks promising.

Calculate t **as in 1c.**

$$t = 3.237$$

Look up the probability as in 1d.

$$df = 16 \qquad p = 0.003$$

The Conclusion: step 1e. We can be 99.7% confident that the treatment improves yield.

A note of caution to advanced t-testers: In the above we have *assumed that the variance is the same in both samples* and that *the samples are independent of each other* because this is usually the case.

However, there are adapted versions of the t-test which should be used when the sample variances are different (e.g., the size of doughnuts in New York and London) or when paired measurements are taken on individuals before and after treatment (e.g., weights during a dieting regime).

9.7 Is the Data from a Normal Distribution?

We have already seen that the t-test relies upon data being "normally distributed." There will be occasions, therefore, when it will be necessary to test that the data does indeed follow a specific distribution. The following test, invented simultaneously by two Russian statisticians, Andrey Nikolaevitch Kolmogorov and Nikolai Vasilevich Smirnov, may be used to check the assumption. It relies upon the comparison of the assumed cumulative distribution and that of the actual sample.

We use the beer consumption data from the provinces, mentioned at the beginning of the chapter and repeated below (in ascending order), as an illustration of the test:

Provinces   6   10   15   15   17   19   20   27   30   33

The mean of the above sample is 19.2 and its standard deviation is 8.613, and we use these values to define the appropriate normal distribution. The cumulative distribution is drawn by plotting the probability of observing a value $\leq x$ from this normal distribution, represented by the continuous curve on the graph in figure 9.2. The cumulative distribution is always a curve that begins at zero and climbs to a value of 1.

The Kolmogorov-Smirnov statistic is given by the largest discrepancy between the two plots. The 10% critical value for the Kolmogorov-Smirnov goodness-of-fit test with $n = 10$ (number of observed points) is 0.3687, which is much larger than our value of 0.163. Therefore we have no evidence to reject the null hypothesis that the data are normally distributed. Large

9.8.2   Contingency Tables: Are Hair and Eye Color Related?

When each member of a sample is classified according to two properties, the observed frequencies can be expressed in a contingency table. As an example consider the following survey in which we have classified individual people according to eye and hair color:

		Hair color			
		Blonde	Brown	Black	Total
Eye	Blue	80	90	40	210
color	Brown	20	210	10	240
	Total	100	300	50	450

We may wish to know whether eye and hair color are related. In order to do this we perform the usual statistician's trick of assuming that they are not (the null hypothesis—H0) and then see if the data are consistent with this assumption. Thus, we need to see if the observed values differ from the expected (which we calculate making the assumption that there is no difference). We are obviously in a situation where the χ^2 test may prove useful.

Now, if we are to use the χ^2 test, how do we calculate the expected values? In this case we have no underlying theory to help, so we must rely on our assumptions and a bit of common sense. Let us look at blue-eyed blondes (who's biased?) as an example. From our table we can calculate

1. the probability of being blonde $= \frac{100}{450}$ (remember that we are assuming that eye color does not influence hair color and that we found 100 blondes in our sample of 450), and similarly

2. the probability of being blue-eyed $= \frac{210}{450}$

Therefore, the expected number of blue-eyed blondes is

$$\frac{100}{450} \times \frac{210}{450} \times 450 = 46.66$$

In order to simplify the calculation we can rearrange this as

$$\frac{\text{total blonde} \times \text{total blue-eyed}}{\text{grand total}}$$

Basically, if we assume that the two factors are independent, we use the row and column totals because they are the best values that we have available. In the same way we can find the expected values for brown-eyed blondes:

$$\frac{\text{total blonde} \times \text{total brown-eyed}}{\text{grand total}}$$

Thus, we can construct a table as follows:

		Hair color			
		Blonde	Brown	Black	Total
Eye	Blue	80 (46.7)	90 (140)	40 (23.3)	210
color	Brown	20 (53.3)	210 (160)	10 (26.7)	240
	Total	100	300	50	450

where the figures in brackets are the expected frequencies.

We calculate the χ^2 statistic using the six observed and expected frequencies from the table.

$$\begin{aligned} \chi^2 &= \frac{(80 - 46.7)^2}{46.7} + \ldots \\ &= 100.45 \end{aligned}$$

The value for degrees of freedom in the case of contingency tables is calculated as follows:

$$df = (rows - 1) \times (cols - 1)$$

because if each row and column has one value missing it is possible to calculate it from the other values and the row and column totals.

We thus look up the value 100.45 with two degrees of freedom. Tables show that this value is much larger than $\chi^2(2)$ at 0.001. We therefore conclude (with confidence in excess of 99.9%) that the null hypothesis must be rejected and that hair and eye color are related in some way.

9.9 The Mann-Whitney Test: Are Two Samples Different?

The Mann-Whitney test is a nonparametric test, equivalent to the classical t-test. It is almost as powerful, but where the t-test is restricted to samples that are assumed to be normally distributed, the Mann-Whitney test makes no assumptions about the underlying distribution and may be applied almost universally. The test involves ranking the values from both groups in a combined list.

Suppose we have 12 items, 6 (n_1) from sample X and 6 (n_2) from sample Y, and a panel has ranked them in the following order:

$$Y \quad X \quad X \quad X \quad Y \quad Y \quad X \quad X \quad Y \quad X \quad Y \quad Y$$

signifying that the best result came from sample Y, the next three from sample X and so on. Note that the observed values needn't be numbers. Our experimental hypothesis is that the two samples are different, but we don't care which is better, so this will be a two-tailed test.

It is possible to calculate the probability of observing any given permutation of the list when both samples are identical, and the Mann-Whitney test allows us to calculate the probability of a given sequence arising from such a situation.

In order to carry out the test, we attach values to each Y according to the number of X's which precede it and sum the results as follows:

Y	X	X	X	Y	Y	X	X	Y	X	Y	Y
0				3	3			5		6	6	total $= 23$

Similarly we could attach values to each X according to how many Y's precede it:

Y	X	X	X	Y	Y	X	X	Y	X	Y	Y
	1	1	1			3	3		4			total $= 13$

The Mann-Whitney statistic, for a two-tailed test such as this, is the smaller of these two totals. This may involve the evaluation of both, but usually it will be obvious which one to calculate. The astute will realize that the sum of these totals is $n_1 \times n_2$ and that the calculation of one total is therefore sufficient. It is interesting to prove this relationship, which I leave as an exercise for the reader. In this case we take the value 13 and compare it with those in table 9.1 with $n_1 = 6$ and $n_2 = 6$.

The null hypothesis will be that there is no difference between the samples.

In order to reject the null hypothesis at the given probability level, the total from our observations should be less than or equal to the critical values in tables 9.1, 9.2, or 9.3. Note that the smaller the Mann-Whitney value, the more significant the result.

In this case we find that there is no justification in believing the samples to be different, since the probability of getting such a sequence by chance is greater than 10%. The critical value here is 7, and our value is 13. We accept the null hypothesis.

If the ranked list had produced the following:

$$X \quad X \quad X \quad X \quad X \quad X \quad Y \quad Y \quad Y \quad Y \quad Y \quad Y$$

we would have no doubts in deciding that samples X and Y are different! The two totals here would be 0 and 36. The value of 0 is less than the 1% value in the tables (2), so we would be justified in rejecting the null hypothesis. This is an obvious case, but it is useful to check that the tables agree with our commonsense feelings!

Example: Is the Frequency of Plant Species Different at Different Locations?

During an ecological survey the number of plants of a given species was counted at various points to the north (N) and south (S) of an estuary.

The counts were made at random sample points using a standard quadrat method, the objective being to show a difference between the north and south locations. The following data were recorded:

North	7	5	9	13	11	7	9	8
South	4	6	12	6	4	3	2	4	3

The null hypothesis is "There is no difference in species abundance." To prove this,we first set up the combined ranked list and count the number of Ns preceding Ss as follows:

2	3	3	4	4	4	5	6	6	7	7	8	9	9	11	12	13
S	S	S	S	S	S	N	S	S	N	N	N	N	N	N	S	N
0	0	0	0	0	0		1	1							7	

The first row is the combined ranked list. The second row tells us where the sample came from, and the third row tells us how many Ns precede each S. Thus we are able to calculate the Mann-Whitney statistic to be 9.

Consult your computer, or a set of tables, with $(n_1 = 9, n_2 = 8)$ to find the critical values, which are 15 at the 5% level and 9 at the 1% level. We therefore conclude that the abundance is different, with a 1% probability of being wrong. (I.e., the likelihood of observing this result by chance, if both sides of the estuary are equally populated, is small enough for us to assert that there must be a real difference.)

Example: Do The Heights of Two Groups Differ?

The heights (cm) of two sample groups of people, B and D, were measured and found to be as follows:

B	165	180	191	160	171	181	169	170
D	169	184	192	171	175	185	186	176	179

We are asked whether the samples differ in height.

For convenience the first digit has been omitted when combining the list and ties have been signalled by the overscores, "=".

60	65	69	69	70	71	71	75	76	79	80	81	84	85	86	91	92
B	B	B	D	B	B	D	D	D	D	B	B	D	D	D	B	D
0	0	.5		1	1.5					5	5				8	

In the case of ties half of each D in the tie is assumed to precede each B. We obtain a total of 21 with $(n_1 = 8, n_2 = 9)$, which is larger than the 10% value (18). Hence, we would accept the null hypothesis and conclude that there is no difference in height between the two groups.

9.10   One-Tailed Tests

So far we have been answering questions of the type "Are the two samples different?" Sometimes we may wish to ask the more specific question "Is this sample better than the other sample?" In this case the calculation proceeds as follows:

Only one sum is calculated—the one that would be expected to be smallest when H1 is true—and this value is looked up in the tables. However, in this case the probability to be used is *half* the value stated in our tables.

Consider the data in the ecological example above, under the experimental hypothesis (H1) that "the species is more abundant on the south side."

If H1 were true we would expect the N counts to occur earlier in the list than the Ss, and therefore we calculate the number of Ss preceding Ns as follows:

$$\begin{array}{ccccccccccccccccc} S & S & S & S & S & S & N & S & S & N & N & N & N & N & N & S & N \\ & & & & & 6 & & & & 8 & 8 & 8 & 8 & 8 & 8 & & 9 \end{array}$$

Mann-Whitney statistic = 63

Clearly this is not significant, and therefore we reject H1. (I hope that in a real situation you would do this without the necessity of performing the calculation, but here I am simply illustrating the method.)

However, if H1 was that "the species is more abundant on the north bank," then we would count the Ns preceding the Ss as before and obtain the statistic 9. This would lead to the conclusion that the species preferred the northern situation, with a confidence level of 99.5%. Here, the experimental hypothesis should be stated before the data is collected; you could well be fooling yourself if you perform the test on data which you have already previewed!

Table 9.1: 10% Critical Values for the Mann-Whitney Two-Sample Test

	2	3	4	5	6	7	8	9	10	11	12	13	14	15
6	-	2	3	5	7									
7	-	2	4	6	8	11								
8	1	3	5	8	10	13	15							
9	1	4	6	9	12	15	18	21						
10	1	4	7	11	14	17	20	24	27					
11	1	5	8	12	16	19	23	27	31	34				
12	2	5	9	13	17	21	26	30	34	38	42			
13	2	6	10	15	19	24	28	33	37	42	47	51		
14	3	7	11	16	21	26	31	36	41	46	51	56	61	
15	3	7	12	18	23	28	33	39	44	50	55	61	66	72

Table 9.2: 5% Critical Values for the Mann-Whitney Two-Sample Test

	2	3	4	5	6	7	8	9	10	11	12	13	14	15
6	-	1	2	3	5									
7	-	1	3	5	6	8								
8	-	2	4	6	8	10	13							
9	-	2	4	7	10	12	15	17						
10	-	3	5	8	11	14	17	20	23					
11	-	3	6	9	13	16	19	23	26	30				
12	1	4	7	11	14	18	22	26	29	33	37			
13	1	4	8	12	16	20	24	28	33	37	41	45		
14	1	5	9	13	17	22	26	31	36	40	45	50	55	
15	1	5	10	14	19	24	29	34	39	44	49	54	59	64

Further Reading

I don't claim to have read every mathematics (or even introductory mathematics) book, but the following is a selection from those that I have found useful. I haven't recommended particular books as must-haves, because individual students respond in different ways to different authors. My advice is to look around and work with several until you find one that is helpful and that you are comfortable with.

Foundation Maths, by Anthony Croft and Robert Davison, Longman, 1995; ISBN 0-582-23185-X. Lots of worked examples and exercises with answers.

Easy Mathematics for Biologists, by Peter C. Foster, Harwood, 1998; ISBN 90-5702-339-3.

Catch Up Maths and Stats: For the Life and Medical Sciences, by M. Harris, G. Taylor, and J. Taylor, Scion, 2005; ISBN 1-904842-10-0. A useful introduction to basic mathematics directed specifically to the life and medical sciences.

Essential Mathematics and Statistics for Science, by Graham Currell and Antony Dowman, Wiley, 2005; ISBN 0-470-02229-9. A good coverage of mathematics relevant to science, especially geared to applications, with good examples based on analyses of experimental results. The latter half, dealing with statistics, is especially useful and clear.

Calculus Made Easy, by Silvanus P. Thompson, Macmillan; ISBN 0-333-07445-9. This is an old book with many editions, but you can still find copies—there are four of mine somewhere! Worth reading, if only for the introduction and the epilogue. But the rest is good too, and the book goes a long way toward doing what the title says!

Matrix Computation for Scientists and Engineers, by Alan Jennings, Wiley, 1977; ISBN 0-471-99421-9. Another old book but with lots of examples and applications making use of matrix algebra. A good book to dip into for inspiration. Jennings' descriptions are clear, informative, and understandable.

Matrix Computations, by Gene H. Golub and Charles F. Van Loan, Johns Hopkins Studies in Mathematical Sciences, Johns Hopkins University Press, 1996; ISBN 0-08018-5413-X, 0-8018-5414-8. An excellent reference book on computing methods in matrix algebra, with clear and comprehensive descriptions.

What Are the Chances? Voodo Deaths, Office Gossip, and Other Adventures in Probability, by Bart K. Holland, Johns Hopkins University Press, 2002. ISBN 0-8018-6941-2. A gentle and interesting introduction to statistics and its application in a wide variety of situations.

Fermat's Last Theorem, by Simon Singh, Harper Collins, 2002; ISBN 1841157910. A good read, taking a tour through the history of mathematics to the solution by Andrew Wiles in 1995. It's fascinating and readable; you will surprise yourself and enjoy the journey.

Index